M. Cresti   A. Tiezzi   (Eds.)

Sexual
Plant Reproduction

With 75 Figures

Springer-Verlag
Berlin Heidelberg NewYork
London Paris Tokyo
Hong Kong Barcelona
Budapest

Prof. Dr. MAURO CRESTI
Prof. Dr. ANTONIO TIEZZI

Dipartimento di Biologia Ambientale,
Universita degli Studi di Siena,
Via P.A. Mattioli, 4,
53100 Siena, Italy

ISBN 3-540-55746-6 Springer-Verlag Berlin Heidelberg NewYork
ISBN 0-387-55746-6 Springer-Verlag NewYork Berlin Heidelberg

Typesetting: Camera-ready by author
31/3145–5 4 3 2 1 0 – Printed on acid-free paper

Preface

The harvest of more than 70% of all crops grown in the European community consists of the reproductive structures of plants. This harvest of reproductive structures depends on successful plant breeding for improved quality, yield and pest resistance. Until recently, plant breeding has often been based on empirical procedures, but this is changing rapidly and a wealth of morphological, physiological and biochemical information on gametogenesis and fertilization has been accumulated. Thus, the detailed genetic and physiological mechanisms of plant reproduction have become increasingly better understood. Indeed, plant cells are particularly suitable for the application of biotechnological methods.

For these reasons, the Biotechnology Action Programme (BAP) of the Commission of the European Communities, during 1985-1989, supported several research projects dealing with fundamental aspects of plant breeding, such as cytoplasmic male sterility, embryogenesis and the physiology of male gametes. This was further complemented by the Biotechnology Research for Innovation, Development and Growth in Europe (BRIDGE) programme which sponsored research on two crucial aspects in this area, i.e. sexual breeding (mechanisms of flower initiation and evocation; differentiation of sex cells; molecular basis of gamete recognition and selection) and plant regeneration (genetics and molecular biology of somatic and zygotic embryogenesis).

Because plant breeding is of crucial importance for the nutritional requirements of man and livestock, the BRIDGE programme organised an intensive course on sexual plant reproduction. This course was directed by Professor Mauro Cresti, one of the leading scientists in the field, and the lectures were given by well known authorities in the different areas. The course was held in Siena, Italy, from November 15 to December 2, 1991 and provided the participants with theoretical and practical information on different aspects of sexual reproduction in higher plants such as: pollen development, female gametophyte development, pollen-pistil interactions, gene expression, genetic engineering, micromanipulation, reproductive physiology of the progamic phase, syngamy, male sterility, ecology of sexual reproduction and biotechnology of fertilization.

This book includes all the lectures presented during the course and available updates of specific scientific information available at the time of printing. It provides, therefore, an overview of the state of the art of plant sexual reproduction.

DGXII EEC (Brussels)   A. Léonard

Introduction

In recent years there has been a tremendous impetus in research on reproductive structures with which there is a growing awareness of the importance of reproductive biology to crop production.

During the late autumn last year, the first intensive course on Sexual Plant Reproduction was conducted here in the Department of Environmental Biology of the University of Siena. The basic concept of the course was to provide young biologists with an update of scientific information by way of theoretical lectures and on hand experience with the most advanced techniques employed in studies on reproductive biology of higher plants.

The occasion brought together a whole spectrum of leading experts in the field from across the whole of Europe and this volume is mainly based on the series of lectures delivered by them during the course in the University Congress Centre of Pontignano. The principle articles included herein provide comprehensive update overview of recent findings and emerging concepts in several major areas of reproductive processes in higher plants.

It is hoped that this book will receive a broader acceptance amongst undergraduate and postgraduate students, research scientists, teachers and those involved in various practical and applied aspects of plant breeding and biotechnology.

We would like to thank all partecipants for contributing to the success of the course and in particular to the staff and students of our Department and the DGXII of the Commission of the European Communities for the financial support.

M Cresti
A Tiezzi

Contents

X

MICROSPORE DERIVED EMBRYOGENESIS

H.G. Dickinson
Department of Plant Sciences
University of Oxford,
South Parks Road,
Oxford OX1 3RB,
U.K.

INTRODUCTION

The period of cellular development that encompasses meiosis and microsporogenesis is of key scientific and commercial interest, but the complex cytological organisation of the anther and the rapid pace of differentiation has made it difficult to investigate by standard molecular methodology. From straightforward RNA-excess DNA/RNA hybridisation Koltunow *et al* (1990) have calculated 25000 genes to be expressed in the anther, of which 10000 may be organ specific. While the majority of these will be involved in the development of anther wall cells, the tapetum and in "post-microspore" pollen development, the expression of a significant proportion of these genes must be responsible for switching into, and maintaining meiotic development, regulating the alternation of generation, and the establishment of the young gametophyte.

A series of elegant experiments in *Arabidopsis* by Bowman *et al* (1989) and in *Antirrhinum* by Carpenter and Coen (1990) have enabled the identification and characterisation of genes which direct the earliest stages of anther development. Batteries of subsidiary genes activated by one or more of these homeotic sequences result in the differentiation of the three cell types of the young anther. The development of the meiocytes and their enveloping tapetal layer then commences with the division of a hypodermal layer of meristematic cells. The principal cellular events of meiosis and microsporogenesis are set out in Figure 1; the meiocytes first become invested by the β 1:3 glucan, callose, and enter meiotic prophase. In many plants these cells are interconnected by cytomictic channels with the result that meiotic prophase takes place in a coenocyte, rather than in individual cells. The tetrad of meiotic products is again invested by callose. Pollen wall formation commences at this stage, taking place beneath the callosic layer. The synthesis of a β 1:3 gluconase by the tapetum releases the microspores which increase considerably in volume and become vacuolate. Reserve accumulation now commences, and the pollen wall "primexine" is augmented by the accretion of sporopollenin precursors from the tapetum. G1, S and G2 stages of the cell cycle take place over a relatively short period and the microspore enters an asymmetrical pollen mitosis I, resulting the formation of the vegetative and generative cells.

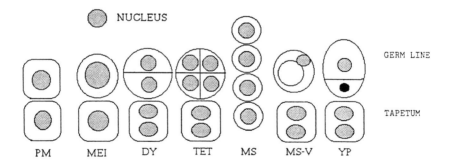

Figure 1.   Schematic diagram of meiosis and microsporogenesis in an angiosperm. See text for full description; stages depicted are PM: premeiosis; MEI: meiotic prophase; DY: dyad; TET: tetrad; MS: young microspore; MS-V: vacuolate microspore; YP: young pollen.

The tapetal cells share some activities with the developing microspores, but instead of entering meiosis, they remain mitotic but often do not complete cytokinesis - generating multinucleate protoplasts. Like the meiocytes the tapetal cells can become linked by cytoplasmic channels to create a coenocyte, but the two tissues themselves never become interconnected. The main activities of the tapetum include protein synthesis, secretion of the ß 1:3 gluconase, synthesis of sporopollenin precursors and orbicules, and the maintenance of the locular fluid in which the young pollen develops.

SWITCHING TO MEIOSIS

Nothing is known of the genes regulating the switch from mitotic to meiotic development in higher plants. In *Saccharomyces* meiosis is held to be suppressed by the product of the ran1+ gene, which is a protein kinase. For example, when cells of *S. pombe* are induced to sporulate by changes in nutritional status, mating type genes are activated, as is a meiotic activator gene mei3+, which encodes a 2.1kDa protein. It is this polypeptide that inhibits the ran1+ product (McLeod and Beach, 1988). Clearly the factors responsible for meiotic induction in plants differ from yeast, but it remains possible that meiosis is repressed in a similar fashion and a different type of switch induces the expression of a meiotic activator gene. Interestingly, heterologous probing of plant genomic DNA with yeast genes indicates the presence of sequences homologous to mei2+ (a yeast meiosis activator) in *Zea*, *Brassica* and *Lycopersicon*. These same species also carry genes with homolology to rme1+ from *S. cerevisiae*, which encodes a protein kinase meiosis inhibitor (P.Howley, A. Bartelesi and H.G.Dickinson, work in progress). While work of this type must be regarded with extreme caution, a similar approach has resulted in the cloning of a number of plant cell cycle genes, such as cdc 2+ and cdc 10+.

Comparison between meiosis in yeast and higher plants is also simplistic because the cell cycles leading up to plant mitosis increase in duration, resulting from an extending S phase (reviewed in Bennett 1984). How this is related to

DNA replication is uncertain, but these is evidence from *Triturus* that replicon length changes in these premeiotic S-phases. Premeiotic interphase is a particularly sensitive period and temperature shock, or treatment with antimicrotubule agents can substantially affect subsequent development, particularly chromosomal pairing, and can even switch the cells "back" to mitotic development (Riley and Flavell, 1977) (See Figure 2). The ph gene in *Triticum* which regulates accuracy of pairing is believed to take its effect in premeiotic interphase (Feldman 1988).

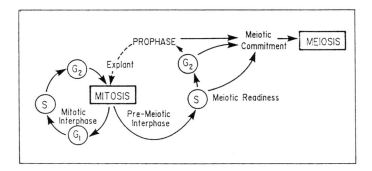

Figure 2.   Events preceding the switch to meiosis (from John, 1990), showing how development of plant meiocytes may be diverted back to mitosis.

THE MEIOTIC PROCESS

The three key events of meiosis are presynaptic alignment of the chromosomes, synapsis (pairing) itself and recombination. From the above it is clear that while events in premeiotic interphase are key to pairing, they are not well understood. The mechanics of the relationship between homologous chromosomes in the two genomes of the diploid cell have only recently been unravelled (Heslop Harrison and Bennett 1984) and accurate data are not yet available for meiosis. Pairing occurs in the leptotene and zygotene stages of meiotic prophase, and there is some evidence that synaptonemal complex (SC) formation commences when homologous telomeres, situated on the nuclear envelope, encounter one another; whether any component of the cytoskeleton is involved is unclear, but microtubules are frequently associated with the nuclear envelope at this point (Sheldon *et al* 1988). The genes encoding SC polypeptides have yet to be identified for higher plants, although a small number of the proteins themselves have been characterised (Heyting *et al* 1988). There are a number of yeast mutants incapable of pairing and recombination; interestingly, when the wild type sequence of the SPO 11 mutant of *S. cerevisiae* (which fails to pair satisfactorily) is used to probe plant genomic DNA, a homologous sequence is clearly identified in *Brassica*, *Zea*, and *Lycopersicon* (Howley, Bartalesi and Dickinson, in preparation).

As in animals, recombination in plants is accompanied by the presence of recombination nodules (Carpenter 1987) which comprise the enzymic machinery involved in DNA nicking, exchange, synthesis and ligation. A number of plant DNA "processing" genes have now been identified and cloned, but none has yet been demonstrated to be specific to meiosis. Although the genetical consequences of meiosis are well understood, the mechanism by which recombination occurs at the level of the DNA molecule remains a matter for speculation; attractive models do exist but conclusive data remains elusive. In a comprehensive series of biochemical and molecular biological investigations Stern and Hotta (1987) studied nucleic acid metabolism during meiotic prophase. Surprisingly, it was discovered that meiosis appeared to be regulated by two phases of late DNA replication. The first occurred during the zygotene stage, and was held to be involved in pairing; it is, however, unlikely that the newly-synthesised DNA is directly involved in pairing, for EM autoradiography shows what DNA synthesis there is at zygotene to be distributed generally throughout the condensing chromosomes (Porter *et al* 1984). The second phase of DNA synthesis took place at the pachytene stage and was essential for crossing over. Stern and Hotta (1987) also described the characterisation of a number of key meiotic polypeptides, including an endonuclease capable of nicking DNA, an unwinding protein, and a protein capable of reassociating DNA - which was absent in asynaptic mutants. Most interestingly, study of the "pachytene-DNA" indicated a mechanism by which recombination was regulated, and recombinational "hot-spots" flagged.

Meiosis differs from mitosis principally in having an extended prophase during which pairing and recombination take place. Thus, as the meiocyte enters the anaphase of meiosis I, development is likely to be regulated by the same cell-cycle genes that control mitosis. Most of the major cell-cycle genes have now been identified in plants, and it would be expected that dephosphorylation of a cdc2/cyclin complex is responsible for the immediate post-prophase events. It is noteworthy that early reports of meiosis in a number of higher plants describe elevated levels of phosphatases at this point (Knox *et al* 1973). How meiosis II is regulated is uncertain for, although it resembles a normal mitotic cell cycle, there is no S-phase suggesting that START must either be absent or modified. It remains possible that the cdc2/cyclin complex never disperses as in normal mitosis, but simply becomes rephosphorylated. There are also differences in the relationship between nuclear division and cytokinesis in higher plant meiosis, with many groups (e.g. the Compositae) undergoing the nuclear divisions of meiosis I and II in a single cytoplasm.

CYTOPLASMIC EVENTS ACCOMPANYING MICROSPOROGENESIS

The earliest observations of pollen development revealed unusual cytological events to accompany microsporogenesis. Guillermond (1924) and later Py (1932) and Wagner (1927) all reported striking changes in the chondriome, and described changes in "granularity". More recent studies have revealed a cycle of organelle dedifferentiation and redifferentiation to accompany meiosis in a number of plants (Dickinson and Heslop Harrison 1970) and, more surprisingly, a depletion of cellular RNA. As soon as the germ line cells become cut off from the investing tapetal tissue by walls of the ß 1:3 glucan, callose, levels of extractable RNA fall dramatically (Mackenzie *et al* 1967). Synthesis of RNA ceases, and pool levels of rRNA and mRNA decrease (Porter *et al*, 1984).

Although unequivocal data are not available, the rate of eradication of RNA appears to exceed normal turnover, indicating the presence of specific nucleases. Certainly meiotic prophase in plants is characterised by the presence of hydrolases (Knox *et al*, 1973), as is also the case in animals, but it is not known whether the RNA is removed by this means. The nucleolar cycle in plant meiosis clearly differs from mitosis, in that rDNA transcripts, produced in late prophase, aggregate in ana- and telophase I and II in some plants to form cytoplasmic nucleoloids (Dickinson and Heslop Harrison, 1977, Sato *et al*, 1991). These disintegrate, repopulating the cytoplasm with ribosomes.

A combination of EM stereometry and autoradiography has revealed the organellar cycle described above also to involve a phase of intense organellar DNA synthesis, followed by rapid division (Bird *et al* 1983). Since this division is not accompanied by further DNA synthesis, the organelles present in maturing pollen contain very few genomic copies (Jones *et al* 1990). Gene expression in these dividing organelles is not well understood, but mitochondria of *Zea* have been shown actively to express genes encoding inner membrane components (F.Moneger and R. Mache, *pers. comm.*), and plastids of *Lilium* commence starch synthesis during the meiotic divisions (Dickinson and Willson, 1983). Mitochondrial associations with the nuclear envelope have also been reported following telophase II in the Compositae; interestingly, present evidence suggests that the role of this interaction is not solely to provide energy for the early haplophase nucleus (Dickinson and Li, 1988).

The cytoskeleton of plant meiocytes again differs from mitotic cells. Prior to the appearance of the spindle microtubules, the prophase nucleus briefly becomes ensheathed by microtubules, some of which appear to connect with the telomeres of the pairing homologues on the inner face of the nuclear envelope (Sheldon *et al* 1988). The division stages of meiosis I and II feature a "mitotic" type of tubulin cytoskeleton, as does prophase II. Following telophase II, radial microtubules are generated from MTOCs on the surface of the nuclear envelope. These extend to the plasma membrane (Dickinson and Sheldon, 1986) and may be involved in the establishment of polarity in the young microspore. A normal cortical microtubular cytoskeleton accompanies intine formation, but there is new evidence that a stage-specific microtubular assembly is responsible for regulating the asymmetry of pollen mitosis I (Zaki and Dickinson 1990, 1991). The actin cytoskeleton of plant meiocytes has also been studied in detail (Dickinson and Sheldon 1990) but no particular activity of microfilaments in the meiotic process has been reported.

GENE EXPRESSION DURING MEIOSIS AND EARLY MICROSPOROGENESIS

Sequences expressed during higher plant meiosis have proved curiously hard to identify and clone. Elegant early work by Bouchard *et al* (1990) resulted in the characterisation of the pLEC series of clones from *Lilium* meiocytes, which have been shown to have strong homology with the hsp 70 family of heat shock proteins, but more recent attempts have so-far only resulted in interesting, but uncharacterised, cDNAs (e.g. Scott *et al* 1991). As with many of the genes expressed in later stages of pollen development, a large number of these sequences are also transcribed in the tapetum and anther wall cells (See Figure 3). Even attempts to make subtractive cDNA libraries from isolated meiocytes have resulted in few meiosis-specific clones (Howley and Dickinson, in preparation), and even some sequences expressed principally in the tapetum (Crossley *et al*, in preparation). Difficulty in cloning genes from meiocytes results not only from the inaccessibility and small number of the cells involved, but also from the very low levels of gene expression during the meiotic process (Howley and Dickinson, in preparation) - presumably caused by the low concentrations of poly A+ RNA present (Porter *et al* 1984).

A study of the sequences cloned during anther development to date (See Figure 3) reveals pollen mitosis I to be a watershed in gene expression, with large numbers of genes being transcribed following this asymmetric division. An analysis of the sequences involved (Willing *et al*, 1988, McCormick, 1991)

Species	Clone names	Sequence similarity	Tissue/cell expression
tomato	LAT51	ascorbate oxidase	pollen
"	LAT52	Zm13, Kunitz trypsin inhibitors (KTi)	pollen
"	LAT56	LAT59, pectate lyases, ragweed allergens, 9612	pollen
"	LAT59	LAT56, pectate lyases, ragweed allergens, 9612	pollen
"	108,92b	_____	tapetum
tobacco	TA29	glycine-rich	tapetum
"	TA56	thiol endopeptidase	anther wall
"	TA36	lipid transfer protein	tapetum
corn	Zm13	LAT52, KTi	pollen
"	Zm58	LAT56, LAT59, pectate lyases	pollen
Brassica	Bp4	lysine- & cysteine-rich	microspore
"	A1-30	range of similarities	meiocyte, microspore & tapetum
Lilium	pLEC series	heat shock proteins	meiocyte
"	pLM9	seed proteins	tapetum
Oenothera	P2	polygalacturonase	pollen
ragweed	AmbaI.1,.2,.3	LAT56, LAT59, pectate lyases, 9612	pollen
Kentucky bluegrass	KBG41,60,30	Lolp1b	pollen
ryegrass	Lolp1b	KBG41,60,30	pollen
sunflower	SF2,SF18	proline-rich	anther wall

suggests that these genes are mainly concerned with the accumulation of reserves, or the synthesis of enzymes required for the growth of the pollen tube in the style, and are thus most likely transcribed in the vegetative cell. Sequences have been cloned encoding enzymes with no obvious function in reserve accumulation or tube growth (e.g. Albani *et al* 1990) but there is no evidence to suggest that they, or indeed any other known genes, are expressed by the generative cell. Interestingly, development during the period prior to pollen mitosis I is particularly labile, and microspores cultured at this stage can develop into embryos. The reasons for this flexibility in development, and the factors causing the switch from a gametophytic to a sporophytic pathway are discussed elsewhere (Dickinson and Zaki, this volume).

Figure 3. (overleaf) A summary table of sequences expressed during pollen development in angiosperms, adapted from McCormick 1991. Source references to all these sequences, except the *Brassica* A series, pLHm9 and the pLEC series are given in McCormick 1991. The *Brassica* sequences are described in Scott *et al* 1991, details of pLM9 will shortly be published (Crossley, Greenland and Dickinson, in preparation), and the pLEC clones have been characterised by Bouchard (1990).

FORMATION OF THE POLLEN WALL

The highly patterned exine of the pollen wall is interesting both genetically and developmentally. Despite the fact that patterning is sporophytically determined, early electron microscopic investigations showed it to be initiated on the microspore surface during the tetrad stage while the products of meiosis are still isolated from the sporophyte by the callose wall (Dickinson, 1970). More recent studies have shown that factors involved in imprinting the pattern on the microspore plasma membrane are generated prior to or during meiotic prophase (reviewed in Sheldon and Dickinson 1983). They may, however, later be modified by the spindle of meiosis II, or by gametophytically-determined processes in the cytoplasm of the microspore. Although the basic architecture is determined whilst the cells are invested by callose, the majority of the sporopollenin of the exine is derived from the tapetum, and accretes onto the primexine surface after dissolution of the callose wall by a tapetally synthesised ß 1:3 gluconase. A number of pollen wall mutants have been reported in *Arabidopsis* and are currently under study.

THE ROLE OF THE TAPETUM

The tapetum is a highly active layer of cells investing the sporogenous tissue. As described above it supplies the major part of the sporopollenin of the exine, but this tissue undoubtably plays an important role in supporting early microsporogenesis. A number of tapetally expressed genes have been characterised (See Figure 3), interestingly many shared with the developing germ line and/or anther wall cells. Homology searches have permitted functions to be assigned of some of these sequences (McCormick, 1991) but many remain unidentified. One class of tapetal genes, exemplified by the pLHM9B clone from *Lilium* (Crossley *et al*, in preparation) and the A9 clone from *Brassica* (Scott *et al* 1991) are highly tapetum specific and encode a polypeptide with homology to seed proteins and enzymic inhibitors. Its function is unknown, although molecules of this type must be involved in establishing the highly-specialised matrix of the locular fluid.

The key role played by the tapetum in pollen development is exemplified by the use of tapetal promoters to drive lethal transgenes in engineered systems of male sterility (e.g. Mariani *et al*, 1990). Indeed, it is possible that the tapetal cells are also the principal site of action of some natural systems of cytoplasmic male sterility (CMS). Evidence from *Petunia* indicates that low levels of DNA synthesis in the tapetum are the first indicators of the CMS phenotype, followed by disruption of the young microspores (Dickinson *et al* 1987). Genetically, CMS results from lesions in mitochondrial DNA, and it is interesting that even prior to these changes in the kinetics of DNA synthesis, lowered ATP/ADP and NADPH/NADP ratios are found in these anthers (Liu and Dickinson, 1989). Rapid DNA synthesis in the tapetum has been reported for a wide range of plants and individual cells - some multinucleate - may contain levels of up to 6C. Preliminary evidence from RFLPs of tapeta suggests that the replication taking place at this time may not involve the entire genome, but consist of preferential amplification of specific sequences (Jones and Dickinson, in preparation).

ACKNOWLEDGEMENTS

The work has been supported by grants from the UK Agriculture and Food Research Council. The author also wishes to thank Ann Rogers for help with the preparation of the manuscript.

REFERENCES

Albani D, Robert LS, Donaldson PA, Altosaar I, Arnison PG, Fabijanski SF (1990) Characterisation of a pollen-specific gene family from *Brassica napus* which is activated during early microspore development. Plant Mol

Biol 15: 605-622

Bennett MD (1984a) Nuclear architecture and its manipulation. In "Gene manipulation in plant improvement", 16th Stadler Genetics Symposium. Ed JP Gustafson. pp 469-502

Bennett MD (1984b) Premeiotic events and meiotic chromosome pairing. Symp Soc Exp Biol 38: 87-121. Eds CW Evans, HG Dickinson

Bird J, Porter EK, Dickinson HG (1983) Events in the cytoplasm during male meiosis in *Lilium*. J Cell Sci 59: 27-42

Bouchard RA (1990) Characterisation of expressed meiotic prophase repeat transcript clones of *Lilium*: meiosis specific expression, relatedness, and affinity to small heat-shock protein genes. Genome 33: 68-79

Bowman JL, Smyth DR, Meyerowitz EM (1989) Genes directing flower development in *Arabidopsis*. Plant Cell 1: 37-52

Carpenter ATC (1987) Gene conversion, recombination nodules and the initiation of meiotic synapsis. BioEssays 6: 232-236

Carpenter R, Coen ES (1990) Floral homeotic mutations produced by transposon-mutagenesis in *Antirrhinum majus*. Genes and Development 4: 1483-1493

Dickinson HG (1970) Ultrastructural aspects of primexine formation in the microspore tetrad of *Lilium longiflorum*. Cytobiologie 1: 437-449

Dickinson HG, Heslop-Harrison J (1970) The behaviour of plastids during meiosis in the microsporocytes of *Lilium longiflorum* Thunb. Cytobios 6: 103-118

Dickinson HG, Heslop-Harrison J (1977) Ribosomes, membranes and organelles during meiosis in angiosperms. Phil Trans Roy Soc Lond B 227: 327-342

Dickinson HG, Li FL (1988) Organelle behaviour during higher plant gametogenesis. In "Division and segregation of organelles". Eds SA Boffey, D Lloyd. Soc Exptl Biol Seminar Series 35: 131-148

Dickinson HG, Liu XC (1987) DNA synthesis and cytoplasmic differentiation in tapetal cells of normal and CMS lines of *Petunia hybrida*. Theor Appl Genet 74; 846-851

Dickinson HG, Sheldon JM (1986) Pollen wall formation in *Lilium*: the effect of chaotropic agents and the organisation of the microtubular cytoskeleton during pattern development. Planta 168: 11-23

Dickinson HG, Sheldon JM (1990) The cell biological basis of exine formation in *Lilium* sp. In "Proc VII Palynology Symposium APLE, University of Granada/CSIC, pp 17-29

Feldman M, Avivi L (1988) Genetic control of bivalent pairing in common wheat: the mode of *Ph1* action. In "Kew Chromosome Conference III". Ed PE Brandham. HMSO (Lond) pp 269-279

Guillermond A (1924) Recherches sur l'evolution du chondriome pendent le development du sac embryonaire et des cellules-mères des grains de pollen dans les Liliacées et la signification des formations ergastoplasmiques. Ann Sci Nat Bot 6: 1-52

Heslop-Harrison JS, Bennett MD (1984) Chromosome order - possible implications for development. J Embryol Exp Morphol 83: Supplement, 51-73

John B (1990) "Meiosis". Developmental and Cell Biology Series 22. Cambridge Univ. Press, Cambridge, UK. 396 pp

Jones K, Crossley S, Dickinson HG (1990) Investigation of gene expression during plant gametogenesis as revealed by *in situ* hybridisation using non-isotopic probes. In *"In situ* hybridisation and the study of development and differentiation". Eds N Harris, D Wilkinson. SEB Seminar Series, CUP (Cambridge), pp 189-203

Knox RB, Heslop-Harrison J, Dickinson HG (1971) Cytoplasmic RNA and enzyme activity during the meiotic prophase in *Cosmos bipinnatus*. In "Pollen, development and physiology". Ed Heslop-Harrison. Butterworth (London)

Koltunow AM, Truettner J, Cox KH, Wallroth M, Goldberg RB (1990) Different temperal and spatial gene expression patterns occur during anther development. Plant Cell 2: 1201-1224

Liu XC, Dickinson HG (1989) Cellular energy levels and their effect on male cell abortion in cytoplasmically male sterile lines of *Petunia hybrida*. Sex Pl Reproduction 2: 167-172

Mackenzie A, Heslop-Harrison J, Dickinson HG (1967) Elimination of ribosomes during meiotic prophase. Nature 215: 997-999

Mariani C, de Beuckeleer M, Truettner J, Leemans J, Goldberg RG (1990) Induction of male sterility in plants by a chimaeric ribonuclease gene. Nature 347: 737-741

McCormick S (1991) Molecular analysis of male gametogenesis in plants. Trends in Genetics 7 (9): 298-303

McLeod M, Beach D (1988) A specific inhibitor of the *ran1+* protein kinase regulates entry into meiosis in *Schizosaccharomyces pombe*. Nature 322: 509-514

Porter EC, Parry D, Bird J, Dickinson HG (1984) Nucleic acid metabolism in the nucleus and cytoplasm of angiosperm meiocytes. In "Controlling events in meiosis". Eds C Evans, HG Dickinson. Company of Biologists (Camb) p 363-369

Py G (1932) Recherches cytologiques sur l'assise nourricière des microspores et les microspores des plants fasculaires. Revue Gén Bot 44: 316-368

Riley R, Flavell RB (1977) A first view of the meiotic process. Phil Trans Roy Soc B 277: 191-199

Sato S, Jones K, de los Dios Alche J, Dickinson HG (1991) Cyhtoplasmic nucleoloids of *Lilium* male reproductive cells contain rDNA transcripts and share features of development with nucleoli. J Cell Sci 100: 109-118

Scott R, Dagless E, Hodge R, Wyatt P, Soufleri I, Draper J (1991) Patterns of gene expression in developing anthers of *Brassica napus*. Plant Mol Biol 17 (2): 195-209

Sheldon JM, Dickinson HG (1983) Determination of patterning in the pollen wall of *Lilium henryi*. J Cell Sci 63: 191-208

Sheldon JM, Willson CE, Dickinson HG (1988) Interaction between the nucleus and cytoskeleton during the pairing stages of male meiosis in higher plants. In "Kew Chromosome Conference III". Ed P Brandham. HMSO pp 27-35

Stern H, Hotta Y (1987) The Biochemistry of Meiosis. In "Meiosis; cell biology" ed Peter B Moens. Academic Press Inc (London, UK) p 303-382

Wagner N (1927) Evolution du chondriome pendent la formation des grains de pollen des angiosperms. Biologie Gén 3: 15-66

Willing RP, Bashe D, Mascarenas IP (1988) An analysis of the quantity and diversity of mRNAs from pollen and shoots of *Zea mays*. Theor Appl Genet 75: 751-753

Willson C, Dickinson HG (1983) Two stages in the differentiation of amyloplasts in the microspores of *Lilium*. Ann Bot 52: 803-810

Zaki MAM, Dickinson HG (1990) Structural changes during the first divisions of embryo resulting from anther and free microspore culture in *Brassica napus*. Protoplasma 156: 149-162

Zaki MAM, Dickinson HG (1991) Microspore derived embryos in *Brassica*; the significance of division symmetry in pollen mitosis I to embryogenic development. Sex Pl Reprod 4: 48-55

GENE EXPRESSION DURING MICROSPOROGENESIS

M.A. Zaki* and H.G. Dickinson

Department of Plant Sciences
University of Oxford,
South Parks Road,
Oxford OX1 3RB,
U.K.

*High Institute of Efficient Productivity,
Zagazig University,
Zagazig,
Egypt.

HISTORY AND INTRODUCTION

Naturally-occurring haploids have been known in the plant kingdom for many years; for example Blackslee *et al* recorded a haploid mutant of *Datura* as long ago as 1922. While it was appreciated that doubled haploids would provide the most effective route for the production of homozygous lines for plant breeding, this approach could not be applied since haploid lines were not available for the majority of crop plants. In any event, for such a strategy to be successful, homozygous lines would have to be produced during the breeding programme - and no method for this existed.

Not surprisingly, some of the earliest attempts to culture plant cells involved pollen grains. Amongst the cell culture pioneers, Tuleke (1953) met with the greatest success when he derived callus from the pollen of *Ginkgo biloba*. There is no record whether this tissue was haploid but, in any event, even if it were, the technology was then not available to regenerate whole plants from tissue.

A reliable method for producing haploids emerged from a chance observation by Guha and Maheshwari who, while investigating meiosis in *Datura in vitro* noted the emergence of growths from cultured anthers. In their classic report of 1964 they identified these outgrowths as pollen-derived embryos, and confirmed their haploid status. In recent years this anther culture methods has been used to generate haploid embryos of *Nicotiana* (Sunderland and Roberts 1977), *Lycopersicum* (Sharp et al 1972) and a range of other crop plants. Even the earliest studies indicated that some species were more amenable to anther culture than others, and for this reason, attempts were made to culture the developing microspores themselves. This has proved a very efficient method for producing haploids in *Brassica* (Lichter, 1982), as well as in species that also respond in anther culture (e.g. *Datura*, Nitsch and Norreel, 1973). The ability to produce large numbers of haploids is of great value to the plant breeder for not only may they be doubled to for homozygous lines, but the haploids themselves - or even the embryos/tissues giving rise to them - may be used in so-called "gametoclonal" selectional programmes (Morrison and Evans, 1987). So far, haploid embryos have been produced from the developing pollen of some 235 species.

Despite this success, haploid lines still cannot reliably be produced from a number of important crop species. In some cases there is no response either to anther or microspore culture, in others only callus is formed during culture, and all attempts to regenerate this into plants have failed. Yet other plants form viable embryos, but these do not survive regeneration. To establish the biological basis of microspore-embryogenesis, a number of investigations have focused on the microspore stage of pollen development, and attempted to explain the extreme developmental lability of this stage. Further, a series of studies have convincingly demonstrated that embryogenesis can take place by a variety of different pathways, depending on the species and, in some cases, the conditions of culture.

PATHWAYS TO EMBRYOGENESIS

In all species investigated, microspore-embryogenesis involves a developmental change at or near the first pollen mitosis. In most cases the typical asymmetrical division fails to occur, less commonly, the division does take place, but the mitotic products undergo very different fates from those followed *in vivo*. Four different developmental pathways (A - D) have been identified so far, and are set out diagrammatically in Figure 1.

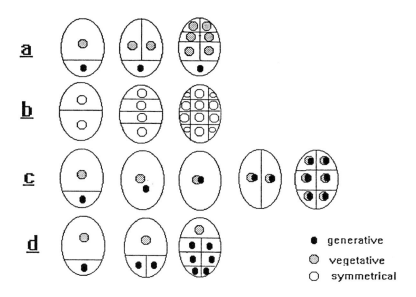

Figure 1a-d. Different pathways to embryogenesis following pollen mitosis I (see text).

In pathway A (Figure 1a), exemplified by *Nicotiana*, the asymmetric cytokinesis of pollen mitosis I does take place, but the nascent vegetative cell continues in division to form the embryoid body. The generative cell eventually becomes necrotic (Sunderland and Wicks, 1971). Pathway B (Figure 1b) involves a symmetrical pollen mitosis I, and division continues to give rise to a globular embryoid; this type of development is characteristic of *Brassica* and members of the graminae (Wilson *et al*, 1978).   Pathway C (Figure 1c) is unusual for it involves the normal asymmetric division, but the products of this mitosis then fuse to form a single "diploid" nucleus, which then continues division to form the embryoid. This pathway is often facultative, with particular plants - for example *Datura* and *Oryza* - apparently able to follow a number of developmental pathways, including C (Sunderland *et al*, 1974). Pathway D (Figure 1d), which has so far only been identified in *Hyoscyamus* (Rhagavan, 1976), resembles pathway A, except in that it is the generative cell that continues division to form the embryoid, rather than the vegetative cell which degenerates.

The route to plant regeneration may also differ, with some species not passing through an embryonic stage, but rather being regenerated from callus, the so-called "direct" and "indirect" pathways respectively (Keller and Armstrong, 1977, Reynolds, 1986). Some plants can be regenerated through both routes, but the direct is preferable for the plant breeder for it is more reliable and less labour-intensive.

REGULATION OF DEVELOPMENT *IN VITRO*

A number of different strategies (for review, see Herbele Bors, 1985) have been developed for inducing microspores, whether contained in the anther or free, to differentiate into embryos. In most cases these involve some form of physical stress or shock, commonly elevated temperatures and high sugar levels. From the earliest studies it emerged that not only was embryogenesis species-dependent, but also heavily reliant on genotype and pretreatment of the donor plants. For example, for good yields of *Brassica* embryos, only a very restricted range of cultivars could be used, and the plants required very high level of photosynthetic irradiance (Lichter, 1982).

The significance of the physiological condition of the cells undergoing induction is exemplified in *Nicotiana*, where Herbele Bors (1980) has identified two types of sporogenous cell; one large, which preferentially develops into an embryoid, and a smaller type which is resistant to induction. Herbele Bors regards the larger type of cell as containing higher levels of "sporophytic" information, which facilitates a reversion to sporophytic development. Such studies have led this group to investigate the possibility of regulating the development of pollen *in vitro* with the aim of developing protocols for inducing sporophytic and gametophytic differentiation. This work has been spectacularly successful with *Nicotiana*, and methods have now been published by Herbele Bors (Benito Moreno *et al*, 1988) and others for the culture of pollen or embryos *in vitro* using uninucleate microspores as starting material.

CELLULAR DEVELOPMENT DURING EMBRYOGENESIS

The cytology of microspore-derived embryogenesis has been studied chiefly in *Nicotiana* and *Brassica*. *Nicotiana*, which generally follows the A pathway, undergoes a normal pollen mitosis I in culture, but in the presence of high sugar levels changes then occur in the vegetative cell. These involve a general dedifferentiation of the cytoplasm (Dunwell and Sunderland 1974a, 1974b and 1975), a progressive loss of starch, and then a commencement of a "somatic" type of cell division. Interestingly, one of the earliest events detected involved the wall of the vegetative cell, which changes from the fibrillar intine to a "normal" cell wall. Other authors have identified cytological features held to be associated with predisposition to embryogenesis but, to date, these have proved difficult to explain in terms of cell development. The later stages of embryogenesis in *Nicotiana* are structurally unexceptional. The cellular mass derived from the vegetative cell develops an epidermis and internal cortex. The cells of these embryoid are undistinguishable from somatic cells of *Nicotiana*, contain normal levels of reserves and invested by fibrillar cellulosic walls traversed by plasmodesmata. The form of the young embryo then assumes the globular and torpedo morphologies characteristic of zygotic development, and finally cotyledons are formed.

Development in *Brassica*, which follows the B pathway, shares a number of features with *Nicotiana*. However, a number of important changes occur prior to pollen mitosis I. Microspores are transferred to culture at the vacuolar stage, with the nucleus apparently appressed to the cell wall. Within a few hours, the vacuolar organisation breaks down and, in the majority of cells, the nucleus moves into a more central position. Pollen mitosis I now takes place, forming two symmetrical daughter cells (See Figure 2a). While these events have been recorded previously (Pechan and Keller, 1988), more recent studies (Zaki and Dickinson 1990) have revealed significant developmental changes in the microspore prior to division. These include the apparent loss of the delicate microtubular cytoskeleton investing the nucleus, the appearance of a thick somatic-type wall and the presence of a population of starch containing plastids investing the centrally-situated nucleus. Pollen mitosis I is clearly a somatic-type division, with a normal cell plate being formed, traversed by plasmodesmata (See Figure 2b) and containing no callose. Large volumes of globular material also characterise the cultured microspore and its immediate mitotic products; the composition of this substance is unknown, but it reacts with stains for saturated lipid, and often appears associated with the vacuolar content.

Figure 2a.　3-cellular pro-embryo of *Brassica napus*, prepared for electron microscopy as described in Zaki and Dickinson (1990). ER: endoplasmic reticulum; S: starch; V: vacuole; G: globular domain; W: cell wall.

x 4,500

Figure 2b.　Primary cell wall (W) formed between the daughter cells of a symmetrical pollen mitosis I in *Brassica napus*. Plasmodesmata (arrows) are easily visible. V: vacuole; M: mitochondrion; G: globular domain.

x 13,100

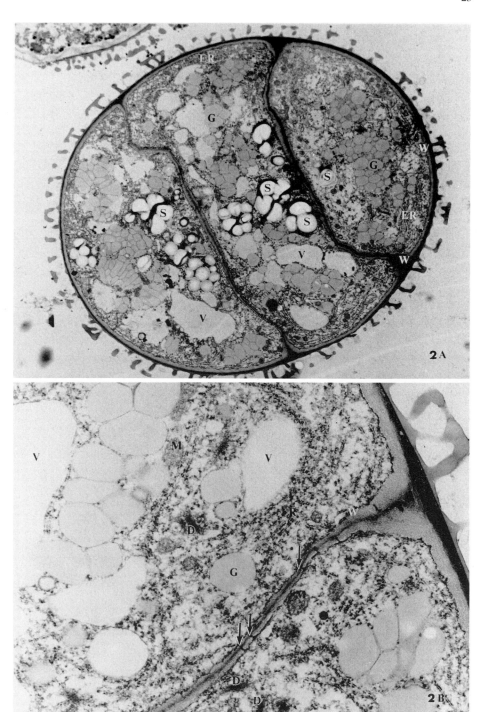

THE MOLECULAR BASIS OF EMBRYOGENESIS

To explain the switch from gametophytic to sporophytic developmental pathways in molecular terms requires two questions to be addressed; one, why is the microspore stage so developmentally labile and, two, at what precise stage does embryonic development commence?

There is accumulating evidence that pollen mitosis I is a crucial stage in gene expression for, with only one or two exceptions (Scott *et al* 1991), all pollen-specific genes cloned so far are expressed from this point onwards. No sequences specific to the immediate post-meiotic stage have yet been characterised, although a number have been cloned in libraries (Scott *et al*, 1991, Howley *et al* 1992). *In vitro* translation experiments reveal the reason for this to be a very low level of gene expression - the lowest recorded in reproductive tissue - during this period (Howley *et al*, 1992), presumably a result of the elimination of the majority of sporophytic message prior to meiosis (Porter *et al*, 1984) and little synthesis of "gametophytic" poly A+ RNA. Thus, if developmental commitment can be considered in terms of synthesis of specific mRNAs and polypeptides, the immediate post-tetrad period in higher plants must, by comparison with other stages, be relatively developmentally labile. It follows that even small changes in gene expression in response to external stimuli could have profound effects on the developmental fates of these cells.

Detailed study of the switch to sporophytic development in *Brassica* indicates that the first deviations from the gametophytic programme involve the loss of the vacuole and the movement of the nucleus to a central position, the appearance of starch containing plastids and the synthesis of a thick "somatic" cell wall (Zaki and Dickinson 1990). Culture involves the transfer of cells into a high-sucrose medium, and the high levels of carbohydrate available may well lead to starch synthesis by plastids. The movement of the nucleus to the mid-point of the cell the formation of the thick wall and, however, clear developmental changes. Studies as long ago as those of Sax (1937) indicate that environmental stress can lead to symmetrical pollen mitosis I. More recently Tanaka and Ito (1981) have shown colchemid to induce symmetrical division in explanted anthers of *Tulipa*.

In an attempt to test the hypothesis that disruption of the tubulin cytoskeleton immediately prior to pollen mitosis I can lead to a symmetrical division, and perhaps thence to embryogenesis, cultured microspores of *Brassica napus* were exposed to pulses of cholchicine treatment, at varying concentrations of the drug (Zaki and Dickinson 1991).

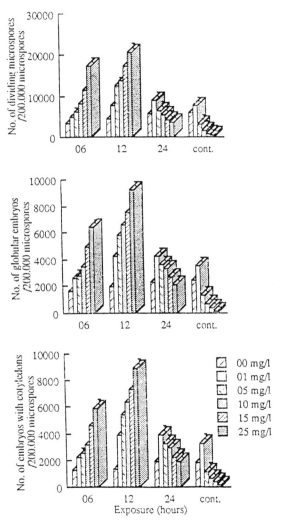

Figure 3.   Development *in vitro* of *Brassica napus* [cv. Topas] microspores following treatment with different levels of colchicine for different periods (from Zaki and Dickinson, 1991).   Top, microspore division, middle, globular embryo formation, and bottom, cotyledonous embryo production.   Vertical bars denote standard errors.

Figure 3 shows typical results from the cultivar Topas, held to be responsive in anther and microspore culture. From the Figure it is evident that a 12h pulse of cholchicine at 25mg.l⁻¹ significantly increases the number of cells dividing symmetrically, and the quantity of embryos produced. These data indicate that the asymmetry of pollen mitosis I relies on a tubulin cytoskeleton present a the end of G2 (See Figure 4), and support for such a structure has recently come from work on cell division of *Nicotiana* BY-2 cell lines (Katsuta *et al*, 1990).

This drug-induced change in division symmetry also appears to have consequences for development. Unfortunately our ignorance of the factors that activate the cascade of gene expression that follows pollen mitosis I makes it difficult to propose a molecular basis for this linkage. It remains possible that the very asymmetry of the division, generating two cells with very different nucleo-cytoplasmic ratios may be sufficient to stimulate patterns of characteristic gene expression. However, the fact that features of somatic development are seen prior to division in the cultured cells (Zaki and Dickinson 1990) suggests that development switches late in G2. The nuclear movement and synthesis of a thick cellulosic wall that take place at this point both involve components of the microtubular cytoskeleton, and it may be that the stress of transfer to culture is sufficient to disrupt the nuclear-positioning microtubules and thereby, or in addition, stimulate normal embryonic development. How the stress of culture could affect microtubules is not clear, but the synthesis of stress response proteins has been reported at this juncture (P.Pechan, *pers.comm*), and such polypeptides are known to interact with the cytoskeleton.

27

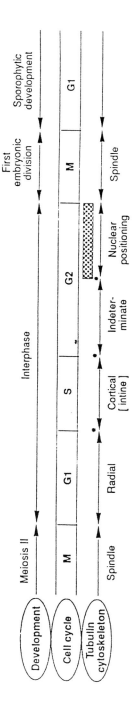

Figure 4. The relationship between development, cell cycle, and the microtubular cytoskeleton during normal pollen development and *in vitro* embryogenesis in *Brassica napus*. Transitions denoted by * must occur but their precise timing has not been established. The crosshatched bar indicates the period when colchicine exerts its maximum effect on embryogenesis (from Zaki and Dickinson, 1991).

ACKNOWLEDGEMENTS

The authors wish to thank the Egyptian Government for financial support to MAZ, and Ann Rogers for help in preparing the manuscript.

REFERENCES

Blackslee AF, Belling J, Farnham ME, Bergner AD (1922) A haploid mutant in the jimson weed, *Datura shamonium*. Science 55: 1433

Dunwell JM, Sunderland N (1974) Pollen ultrastructure in anther cultures of *Nicotiana tabacum* I. Early stages of culture. J Exptl Bot 25: 352-361

Dunwell JM, Sunderland N (1974) Pollen ultrastructure in anther cultures of *Nicotiana tabacum* II. Changes associated with embryogenesis. J Exptl Bot 25: 363-373

Dunwell JM, Sunderland N (1975) Pollen ultrastructure in anther cultures of *Nicotiana tabacum* III. The first sporophytic division. J Exptl Bot 26: 240-252

Guha S, Maheshwari SC (1964) *In vitro* production of embryos from anthers of *Datura*. Nature 204: 497

Herbele-Bors E (1985) *In vitro* haploid formation from pollen: a critical review. Theor Appl Genet 71: 361-374

Herbele-Bors E, Reinert J (1980) Isolation pollen culture and pollen dimorphism. Naturwissenschaften 67: 311

Katsuta J, Hashiguchi Y, Shibaora H (1990) The role of the cytoskeleton in positioning the nucleus in premotitic tobacco BY-2 cells. J Cell Sci 95: 413-422

Keller WA, Armstrong KC (1977) Embryogenesis and plant regeneration in *Brassica napus* anther culture. Can J Bot 66: 1383-1388

Lichter R (1982) Induction of haploid plants from isolatd pollen of *Brassica napus*. Z Pflanzenphysiol 105: 427-434

Morrison RA, Evans DA (1987) Gametoclonal variation. Plant Breeding Rev 5: 359-392

Nitsch C, Norreel B (1973) Effect d'un choc thermique sur le pouvoir embryogene du pollen de *Datura innoxia* cultive dans l'anthere ou isole de l'anthere. C R Acad Sci (Paris) 276D: 303-306

Pechan P, Keller WA (1988) Identification of potentially embryogenic microspores in *Brassica napus*. Physiologia Plantarum 74:377-384

Reynolds TL (1986) Pollen embryogenesis in anther culture of *Solanum carolinense*. Plant Cell Rep 5: 273-275

Rhagavan V (1976) Role of the generative cell in androgenesis in henbane. Science 191: 388-389

Sax K (1937) Effect of variations in temperature on nuclear division in *Tradescantia*. Am J Bot 24: 218-225

Scott R, Daglass E, Hodge R, Paul W, Soufleri I, Draper J (1991) Patterns of gene expression in developing anthers of *Brassica napus*. Plant Mol Biol 17 (2): 195-209

Sharp WR, Raskin RS, Sommer HE (1972) The use of tissue culture in the development of haploid clones in tomato. Planta 104: 357-361

Sunderland N, Collins GB, Dunwell JM (1974) The role of nuclear fusion in pollen embryogenesis of *Datura innoxia*. Planta 117: 227-241

Sunderland N, Roberts M (1977) New approach to pollen culture. Nature 270: 236-238

Sunderland N, Wicks FM (1971) Embryoid formation in pollen grains of *Nicotiana tabacum* J Expt Bot 22: 213-226

Tanaka I, Ito M (1981) Control of division in explanted microspore of *Tulipa gesneriana*. Protoplasma 108: 329-340

Tuleke WR (1953) A tissue derived from the pollen of *Ginkgo biloba*. Science 117: 599-600

Wilson HM, Max G, Foroughi-Wehr B (1978) Early microspore divisions and subsequent formation of microspore calluses at high frequency in anthers of *Hordeum vulgare*. J Exptl Bot 29: 227-238

Zaki MAM, Dickinson HG (1990) Structural changes during the first divisions of embryos resulting from anther and free microspore culture in *Brassica napus*. Protoplasma 156: 149-162

Zaki MAM, Dickinson HG (1991) Microspore-derived embryo in *Brassica*; the significance of division symmetry in pollen mitosis I to embryogenic development. Sex Pl Reprod 4: 48-55

GENE EXPRESSION DURING POLLEN DEVELOPMENT

C. Frova and M.E. Pè
Department of Genetics and Microbiology
University of Milano
Via Celoria 26, 20133 Milano
ITALY

Introduction

Because of allelic segregation at meiosis, post meiotic gene expression can result in considerable genetic variability in the pollen population even from single plants, and thus offers opportunities for selection among male gametophytes. Due to the haploid state and the large population size, male gametophytic selection (MGS) can be extremely efficient and represent an important factor in the high evolution rate of Angiosperms. Moreover, if used in a controlled way, it can be a powerful tool for manipulating the genetic makeup of many useful plant species. Prerequisites for this to occur are: i) that a considerable amount of genes are expressed postmeiotically, thus producing a large number of pollen phenotypes, and ii) that many of these genes are expressed also in the sporophytic phase. Only in this latter case MGS is expected to exert a significant effect on the resulting sporophytic generation.

Another important point regards the timing of gene expression during the male gametophytic phase, which is characterized by several differentiation steps and different functional stages. Excellent reviews (Mascarenhas 1975, Heslop-Harrison 1987) provide a detailed description of the male gametophyte development. Here only those aspects relevant to gene expression will be summarized.

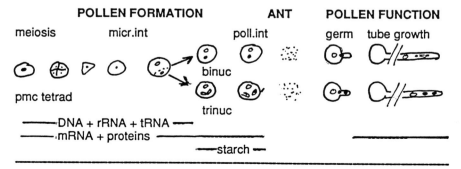

Fig.1 Developmental stages of Angiosperm pollen

During pollen formation (see FIG.1) metabolic activity is high in both microspore and pollen interphase. Microspore mitosis is a crucial developmental step that marks the transition from undifferentiated microspore to two well defined cell types, the vegetative and the generative cells, that follow different developmental programs. An active synthesis and storage of molecules (starch, mRNAs, proteins) which will be utilized later during germination, characterizes the pollen interphase. At dehiscence, on the other hand, pollen is metabolically quiescent.

Activity is resumed upon deposition onto a receptive stigma. The first steps of pollen function (germination) rely mostly upon mRNAs and proteins accumulated during pollen formation. In later stages, however, *de novo* transcription and translation of pollen genes occur. Tube elongation and tube growth rate are clearly dependent of active pollen-style interactions which are, at least part, likely associated with pollen gene expression, although specific genes involved in the process have been so far identified only in few cases.(Bianchi and Lorenzoni 1975, Ebert et al. 1989).

The events briefly described above raise two important questions: 1) is the development of the male gametophyte genetically dependent of the sporophyte or of post meiotic transcription of its own genes? And 2) if the second hypothesis is true, are the changes in pollen characteristics through development the result of differential (stage specific) pollen gene expression?

In the next section approaches that can serve to answer these questions are illustrated and the results discussed.

Analysis of pollen gene expression

Two basic approaches can be utilized to detect gametophytic gene expression:
1) A biochemical approach, consisting in the analysis of gene products, either a) proteins (enzymatic or not), or b) mRNAs
2) A genetical approach, that is the analysis of mutations expressed in the male gametophytic phase.

Each of the above methods presents specific advantages and limits. The choice depends on the kind of information required in each case, but in general the integration of more approaches will yield the most informative results.

1a) Protein analysis. Two classes of proteins can be chosen: multimeric enzymes and proteins the synthesis of which is induced at the transcriptional level by specific stimuli.

In the case of multimeric enzymes, the analysis is based on their electrophoretic profiles. In plants heterozygous for the electrophoretic mobility of a given enzyme (Fast/Slow), the expected zymograms of sporophytic tissues are multibanded, since both homo- and heteromultimers are formed in each 2n cell. In the simplest case of a dimeric enzyme, three bands, SS, FF, SF are formed (FIG.2). The pollen of that same plant is expected to display the identical multibanded pattern only when the genetic control of the enzyme is entirely sporophytic, i.e. if the enzyme itself or the relative mRNA molecules are inherited from the diploid mother plant. If, on the other hand, transcription occurs after meiosis, each haploid grain will produce only one type of homodimer (FF or SS), depending on the allele it carries, and the expected zymogram of a pollen population should show only two (parental) bands.

SPOROPHYTE			F1 GAMETOPHYTE	
			diploid transcr	haploid transcr
SS	FF	F1		
—		—	—	—
	—		—	
—	—		—	—

Fig. 2. Expected zymograms for a dimeric enzyme in the sporophyte and in pollen

The method presents several advantages. It discriminates very accurately between haploid and diploid control of pollen enzymatic activities; it refers to single genes with known functions; it is easy, rapid and inexpensive; it requires little material (3-5 mgs) and therefore is suitable for the analysis of very early stages, and in general to uncover temporal specificity of expression; in post pollination studies it identifies the maternal or paternal origin of activity in pollinated silks. On the other hand, i) it does not cover a large part of the potential post meiotic gene expression, such as regulatory genes or non enzyme coding structural genes, and ii) pollen specific gene expression is rarely detected, because informations about enzymatic functions to be analyzed are generally derived from data on sporophytic tissues.

Anyway, by this method male gametophyte gene expression has been explored in several plant species (Ottaviano and Mulcahy 1989 and Tab. 1), although only in maize was the analysis extended to all stages of pollen development (Frova et al. 1987). In each case, between 20 and 40 isozymes were considered and the overall estimate of haploid and haplo-diploid gene expression made. The results (Tab. 1 and 2) indicate that: i) a large proportion of the genome is post-meiotically expressed; ii) gametophytic-sporophytic genetic overlap is conspicuous; iii) at least 50% of pollen expressed genes show temporal regulation (Frova et al 1987, Mascarenhas 1990).

Very similar results have been obtained by the analysis of transcriptionally induced proteins. In this case the mere presence of the gene product, i.e. a polypeptide band on an electrophoretic gel, can be considered good evidence of the expression of the encoding gene. This approach offers the possibility of exploring a wider part of the plant genome and is non selective with regard to pollen specific genes. It is not just a coincidence that in maize 3 out of 11 heat shock proteins (HSPs) resulted male gametophyte specific versus 1 out of 34 isozymes. Again single gene expression is analyzed, although the function may be unknown. So far this kind of analysis has been performed only for maize and sorghum HSPs (Frova et al 1989, 1991), but its potential application to other plant species and/or other inducible genetic systems (cold, light, salt, chemicals etc.) is large.

1b) mRNA analysis. The studies on the mRNA populations are a very effective approach to obtain quantitative estimates of gene expression based on a large number of genes. The analysis of the hybridization kinetics between cDNA and poly(A)$^+$RNA (R_0t curves) from mature pollen has shown that more than 20,000 different mRNAs are present in *Tradescantia paludosa* and *Zea mays*, compared to the 30,000 transcribed in sporophytic tissues (Willing and Mascarenhas, 1984, Willing et al., 1988). Furthermore the dissection of the R_0t curves has allowed the identification in maize of 3 classes of abundancy in the mRNA population: *i-* mRNA very abundant, accounting for 35% of the total mRNA; *ii-* a medium abundant mRNA sequences that represent about 49% of the mRNA; *iii-* rare mRNA made up by 17,000 different mRNA present in about 200 copies per cell.

These results reveal the extreme complexity of gene expression in the male gametophyte, although this kind of analysis is not able to discriminate between pre- and post-meiotic mRNAs. Similar experiments based on heterologous hybridization between cDNA from pollen and mRNAs from sporophytic tissues, and reciprocal, demonstrated that at least 65% of the genes show haplo-diploid expression, confirming on a very large base that the genetic overlap between the two phases of the plant cycle is quite large (Willing & Mascarenhas, 1984; Willing et al., 1988).

TAB.1 - ESTIMATES OF THE PERCENTAGES OF GENES EXPRESSED IN POLLEN AND IN THE SPOROPHYTE

Expressed in Pollen only	Expressed in both Pollen and Sporophyte	Expressed in Sporophyte only	Assay	References
3%	58%	39%	isozyme (tomato)	Tanksley et al., 1981
6%	73%	21%	isozyme (corn)	Sari et al.,1986
11-17%	74-78%	8-10%	isozyme (Populus)	Rajora et al., 1986
10%	63%	31%	isozyme (barley)	Pedersen et al., 1987
15%	54%	31%	mRNA (Tradescantia)	Willing et al., 1984
	65%		mRNA (corn)	Willing et al., 1988
	(65%)	35%	de mutants (corn)	Ottaviano et al., 1988

The figure in brackets indicates the percentage of sporophytically expressed genes expressed in pollen (modified from Ottaviano and Mulcahy, 1989).

Molecular biology has provided the tools for the identification and isolation of genes specifically or preferentially expressed in pollen and for their fine characterization, which is a prerequisite for an efficient manipulation of the reproductive system. Pioneer were the works carried on in R. Goldberg's and J. Mascarenhas' laboratories, who first isolated anther and/or pollen specific genes from recombinant cDNA libraries (Goldberg, 1988; Stinson et al., 1987). Following this strategy anther- and pollen-specific genes have been isolated from a number of species. These clones can be very useful to study gene regulation during microsporogenesis or anther development (for an extensive discussion, see Mascarenhas 1989, 1990, Koltunow 1990). In fact, two different classes of mRNA were identified by using a few pollen-specific genes of *T. paludosa* and

Zea mays. The first one represents those mRNAs synthesized right after microspore mitosis and whose synthesis steadily increases through microsporogenesis; the second class accounts for the so-called early genes whose transcription begins after meiosis, reaches a maximum during late interphase and then declines (Mascarenhas et al., 1986 and Tab.3). Moreover, the comparison of the sequencing data (genomic clones) derived from a number of genes, and the use of reporter genes and plant transformation strategies, recently led to the identification of molecular structures responsible for temporal and spatial gene expression during male gametophytic development. S. McCormick and coll. were able to identify cis- regulatory elements in the promoter region of 3 genes from tomato preferentially expressed during microsporogenesis and described a modular organization of different regulatory elements that coordinate gene expression in different phases of the plant cycle and in different tissues (Twell et al., 1990; 1991).

2) Mutant analysis The major drawback of the previously described approaches is that it is often difficult to understand the function of the isolated genes. A more direct approach is based on the analysis of mutants affecting microsporogenesis and/or pollen function. Comprehensive reviews on the importance of mutant analysis for the dissection of the genetic program controlling the gametophytic generation have been published (Ottaviano et al 1991). So far, however, the available mutants are limited in number and poorly characterized at the biochemical and molecular level.

Most of the mutations affecting pollen viability are controlled by the sporophyte. One of the few known cases of gametophytic control of such characters is Rf_3, a nuclear gene that restores fertility in cms-S cytoplasm male sterile maize genotypes. Rf_3/rf_3 plants segregate normal and sterile pollen grains with a 1:1 ratio (Laughnan & Gabay, 1983).

Another class of mutants particularly useful either for the dissection of the genetic program that regulates the male gametophyte or for the genetic manipulation of important sporophytic traits by MGS are the sporophytic-gametophytic mutants affecting pollen development and/or function. In maize a set of 34 defective endosperm (*de*) mutants which influence endosperm development causing a reduction in kernel dimension was tested for gametophytic effect. Simple statistical analysis of the distortion from Mendelian segregation ratios in de/+ heterozygous plants uncover the different ability in accomplishing a successful fertilization between normal pollen and pollen carrying the *de* mutation. Three classes of defective endosperm with gametophytic effect (*de-ga*) mutants were revealed: *i-* *de-ga* mutants that affect microsporogenesis; *ii-* mutants showing effects on both pollen development and function; *iii-* mutants that influence pollen tube growth (Ottaviano et al., 1988).

Some of these mutant genes have been recently isolated by gene tagging (Colombo et al 1991), and their molecular characterization is now possible.

TAB.2 - TEMPORAL GENE EXPRESSION IN MAIZE POLLEN

Pollen formation		Pollen function	
ENZYMES			
Adh-1, β-Glu-2[b], Got-1[a], Got-2, Got-3, Cat-3, Cat-4	Adh-1, β-Glu-2[b], Got-1, Got-2, Got-3, Cat-1, Cat-4	Adh-1, Got-1, Got-2, Got-3, Cat-1	Got-1
HSPS (kD)			
102,84,74[b],72 56[b],46,25,18,17	72,64[b]		
mRNA			
Actin	Actin	N.T.	N.T.
	pTpc 44[b]	pTpc44[b]	pTpc44[b]
	pTpc 70[b]	pTpc30[b]	pTpc30[b]
	pZmc 30[b]	N.T.	N.T.
	pZmc 13[b]	N.T.	N.T.
	pLat 52[b]	N.T.	N.T.
	pLat 56[b]	N.T.	N.T.
MUTANTS	de B-116	de B-116	
	de B-1		de B-1
			de B-18

a= sporophytic control
b= pollen specific genes

Discussion

All the results obtained in the past two decades with different experimental approaches indicate that pollen development and pollen function are under the control of a very complex genetic system, ruled by both the sporophyte and the gametophyte. Besides, the amount of haplo-diploid gene expression is relevant , thus allowing to introduce MGS along with the traditional plant breeding systems in order to increase selection efficiency for complex agronomical traits. This strategy has so far produced successful results with regard to cold and salt stress and to herbicides (for a review see Ottaviano et al., 1991). Furthermore the fine analysis of the structure of a few genes expressed exclusively during microsporogenesis has led to the identification of specific regulatory DNA sequences. These informations enabled Mariani and coll. (1990) to directly manipulate the plant reproductive system. They produced in Brassica male sterile plants by transforming with a suicidal gene under the control of an anther-specific promoter from tobacco.

Unlike microsporogenesis, the genetic control of pollen germination and pollen tube growth remains substantially obscure. In particular, besides the self-incompatibility mechanisms, very little is known about the complex interactions taking place between pollen and pistil tissues. Detailed analysis of available mutants and the induction of new ones could play an important role in elucidating these phases.

References

Bianchi A., Lorenzoni C. (1975) Gametophytic factors in *Zea mays*. In: "Gamete Competition in Plants and Animals" (Mulcahy D.L. ed.) pp. 257-264, Elsevier, Amsterdam.

Colombo L., Pè M.E., Binelli G., Ottaviano E. (1991) Induction and isolation of maize *defective endosperm* mutants with gametophytic effect by transposon tagging. In: "Angiosperm Pollen and Ovules: Basic and Applied Aspects" (D.L. Mulcahy, E. Ottaviano and M. Sari-Gorla, eds.), Springer, N.Y., in press.

Ebert P.R., Anderson M.A., Bernatzky R., Altschuler M., Clarke A.E. (1989) Genetic polymorphism of self-incompatibility in flowering plants. Cell **56**: 255-262.

Frova C. Binelli G., Ottaviano E. (1987) Isozyme and hsp gene expression during male gametophyte development in maize. In: "Isozymes: Genetics, Development and Evolution", (Rattazzi M.C., Scandalios J.G. and Whitt G.S. eds), pp 97-120, Alan Liss Inc., N.Y..

Frova C., Taramino G., Binelli G. (1989) Heat-shock proteins during pollen development in maize. Dev. Genet.**10**: 324-332.

Frova C., Taramino G., Ottaviano E. (1991) Sporophytic and gametophytic heat shock proteins synthesis in *Sorghum bicolor*. Plant Sci. **73**: 35-44.

Goldberg R.B. (1988) Plants: novel developmental processes. Science, **240**, 1460-1466.

Heslop-Harrison J. (1987) Pollen germination and pollen tube growth. Rev. Cytology **107**: 1-70.

Koltunow A.M., Truettner J., Cox K.H., Wallroth M., Goldberg R.B. (1990) Different temporal and spatial gene expression patterns occur during anther development. The Plant Cell, **2**, 1201-1224.

Laughnan J.R., Gabay-Laughnan S.J. (1983) Cytoplasmic male sterility in maize. Ann. Rev. Genet. **17**: 27-48.

Mariani C., De Beuckeleer M., Truettner J., Leemans J., Goldberg R.B. (1990) Induction of male sterility in plants by a chimaeric ribonuclease gene. Nature, **347**, 737-741.

Mascarenhas J.P. (1975) The biochemistry of Angiosperm pollen development. Bot. Rev. 41: 259-314.

Mascarenhas J.P. (1989) The male gametophyte of flowering plants. The Plant Cell **1**: 657-664.

Mascarenhas J.P. (1990) Gene activity during pollen development. Ann. Rev. Plant Physiol. Plant Mol. Biol. **41**: 317-338.

Ottaviano E., Pè M.E., Binelli G. (1991) Genetic manipulation of male gametophytic generation in higher plants. In Subcellular Biochemistry, Vol. **17**: Plant Genetic Engineering. Biswas B.B., Harris J.R. (eds) Plenum Press, N.Y., pp. 107-142.

Ottaviano E., Mulcahy D.L. (1989) Genetics of Angiosperm Pollen. Advances in Genetics **26**: 1-64.

Ottaviano E., Petroni D., Pè M. E. (1988) Gamethophytic expression of genes controlling endosperm development in maize. Theor. Appl. Genet. **75**: 252-258.

Stinson J.R., Eisenberg A.J., Willing R.P., Pè M.E., Hanson D.D., Mascarenhas J.P. (1987) Genes expressed in the male gametophyte of flowering plants and their isolation. Plant Physiol., **83**, 442-447.

Twell D., Yamaguchi J., McCormick S. (1990) Pollen specific gene expression in transgenic plants: coordinate regulation of two different tomato gene promoters during microsporogenesis. development **109**: 705-713.

Twell D., Yamaguchi J., Wing R.A., Ushiba J., McCormick S. (1991) Promoter analysis of genes that are coordinately expressed during pollen development reveals pollen-specific enhancer sequences and shared regulatory elements. Genes and Devel. **5**: 496-507.

Willing R.P., Mascarenhas J.P. (1984) Analysis of the complexity and diversity of mRNAs from pollen and shoots of *Tradescantia*. Plant Physiol. **75**: 865-868.

Willing R.P., Bashe D., Mascarenhas J.P. (1988) An analysis of the quantity and diversity of messenger RNAs from pollen and shoots of *Zea mays*. Theor. Appl. Genet. **75**: 751-753.

Cytological Techniques to Assess Pollen Quality

J.S. Heslop-Harrison
Karyobiology Group
Department of Cell Biology
JI Centre for Plant Science
Norwich NR4 7UJ England

The Aim of Assessing Pollen Quality

Viable pollen is pollen that is competent to deliver two male gametes to the embryo sac. Methods to assess pollen quality aim to find the capacity of pollen - individual grains or members of a population - to deliver gametes. Pollen quality is normally measured 1) by scoring seed set in plants fertilized with a particular pollen sample, 2) by cytochemical staining of the grains, or 3) by looking at germination of the pollen *in vitro* or on styles. Table 1 includes the sequence of events that a viable pollen grain must complete. A failure at any stage gives male sterility, since the pollen is then unable to deliver the gametes.

Table 1. The sequence of events in an angiosperm pollen grain's formation and life.

A pollen mother cell must:
    go through meiosis,
    divide and
    differentiate into pollen grains.
A pollen grain must:
    dehisce, and perhaps need a maturation period,
    attach to a stigma,
    hydrate,
    germinate and
    produce a pollen tube.
A pollen tube must:
    penetrate the stigma,
    enter the pollen tube transmitting tract,
    the generative nucleus must divide (in the 70% of families where the pollen is binucleate when shed),
    the tube must grow through the style to the ovule and
    enter the embryo sac and release the gametes.

The Need to Assess Pollen Quality

Pollen quality must be assessed to find plant fertility, to monitor pollen state during storage, in ecological or taxonomic studies and in research on pollen biochemistry, genetics and stigma interactions, incompatibility systems and fertilization (see Stanley and Linskens 1974; Heslop-Harrison, Heslop-Harrison and Shivanna 1984). The tests described here allow measurement of pollen viability by examining and measuring cellular features characteristic of living cells, and they indicate reasons for inviability. Knowledge the nature of inviability can often enable manipulation to overcome the barrier (see, e.g., van Tuly, this volume) although factors influencing pollen viability have proved elusive to understand and control (Knox, Williams and Dumas 1986).

Pollen viability is important in agriculture and for plant breeders since pollen must be viable at the time of pollination for seed (or fruit) set to occur. In apple production, boron deficiency can cause low pollen germinability, and hence poor fruit set; application of boron can correct the problem. Recently, a new variety of wheat in the UK, Moulin, failed in its first year of release because a period of cool, dull and wet weather during microsporogenesis caused poor pollen production. In other contexts, pollen may be stored for germplasm conservation, to make hybrids between plants that flower at different times or places, or for later use in hybridization programmes, and the quality must be monitored. Many economically important plants are propagated vegetatively by bulbs, ramets, tubers or cuttings - potatoes, rubber trees, and lilies are examples. But sexual hybridization is still required for breeding new varieties. Studies of pollen quality are particularly important because the crops often show low fertility and are hence difficult to intercross. Plant breeders use wide hybrids between different species or even genera to transfer alien genes into major crop plants: in wheat, such chromosome engineering is now the only method of genetic manipulation with new genes (see, e.g., Gupta and Tsuchiya 1991). However, the barriers to fertility both in making the hybrids and in the resultant hybrids usually require careful investigation.

Pollen State

Pollen is living, and, like any living organism, its behaviour and survival are influenced by both environment and genotype. Therefore, pollen quality must be measured under defined conditions, and thought about experimental and plant growth conditions is always required. In experimental treatments, possible variables must be carefully considered, and controls used. Pollen production and quality vary from hour to hour, day to day and season to season. Many species have strong diurnal rhythms: in the summer, anther dehiscence may occur at 5 am, and pollen collected 12 h later may be inviable. The nutrition of the parent plant can also have large effects on pollen viability (and may affect results from the *in vitro* germination tests discussed below more than growth on the stigma).

Some pollen will show no viability in tests unless it is correctly preconditioned by, for example, leaving in a humid atmosphere before testing. For most species, the requirements for such preconditioning or post-maturation are not established, so cytological methods to estimate viability may give an unrealistic estimate of quality.

Safety of Tests

Pollen is a severe allergen, and hence is a hazardous substance in the laboratory. Plants with wind dispersed pollen should be kept in enclosed areas and a dust mask used if you are working closely with such plants. Wash away spilt pollen from working areas and from your hands and face. Allergies can be induced by repeated contact with pollen, so insensitive people must be particularly careful.

Most of the chemicals used in pollen tests are of low hazard, and good laboratory practice (GLP) can be used to handle them at the low concentrations involved in the tests, although fluorescein diacetate and DAPI are toxic, so they should be dispensed in a fume cupboard before dilution. The chemicals used are not environmental hazards in small quantities and contain no heavy metals, so they can generally be disposed of to mains drainage or with domestic refuse.

Microscopy

Light microscopy is one of the most important techniques for examining pollen quality since it allows clear observation of pollen morphology. It is useful to examine pollen before carrying out any of the other tests described here to monitor its condition. Staining methods are discussed below, but direct examination of anthers and pollen will confirm that grains are present in anthers, and show undifferentiated or grossly shrunken grains. Most pollen samples will have a few such inviable grains, but the presence of many aborted grains indicates substantial infertility. The causes may include genetic sterility of hybrids, or severe environmental stress. Since microscopy is quick, good pollen can be immediately used for pollinations, and field examination is possible.

Pollen can be observed as shed from the anther under a dissecting microscope, dry at low power under a transmitted light microscope, or at higher powers when mounted in a medium of suitable osmotic strength (typically 10-20% sucrose to prevent bursting), or suspended in an organic solvent such as acetone. Electron microscopy, although not a method for "scoring" pollen viability may be vital to investigate reasons for pollen sterility (e.g. Mogenson and Ladyman 1989).

Seed Set

Measurement of seed set examines the capability of pollen to fertilize a given plant, and in some species provides a standard against which to test other methods. Every step listed in table 1 must be successful, so the seed set test involves assessment of not only the state of the pollen, but also that of the female parent, and of the compatibility of the pollen with the female parent.

Clean (unpollinated) receptive, female stigmata are pollinated lightly with the test pollen. Flowers with mature, unpollinated stigmata can be used, but some species require female inflorescences to be emasculated or bagged to prevent pollination until the stigmata are fully receptive. The pollen for testing is applied to the stigma with some care: too much pollen may prevent hydration, while use of too little may be unsuccessful because of the mentor effect, where several grains must be together before they will germinate and grow successfully. In some species, rough handling of the stigma will cause it to become unreceptive, while in other species, such as *Vicia faba*, scarification is essential (see chapter on The Stigma). The success of pollination can be scored after a few days by looking for fertilized ovaries, and seeds can be counted after ripening, usually 1 to 3 months after pollination.

The test uses the pollen for fertilization, and hence the genes from the pollen can be analysed in later generations. Very low frequencies of viable pollen may be detected, and field scale tests may be carried out. The test examines every step of the compatibility between the pollen and female parent, and can detect nuclear damage to the pollen (from radiation). However, it is slow, percentages of good pollen grains cannot be obtained easily, and female fertility and compatibility are also tested. Contamination, parthenocarpy and parthenogenesis may be problems, so genetic markers may be required to confirm the male parent of seeds. Finally, artificial pollination is not routinely successful for some plants, including economically important crops - rubber and perennial *Arachis* (Lu, Mayer & Pickersgill 1990).

Pollen 'Staining' Tests
Fluorochromatic reaction (FCR test)
This is among the best and most widely used tests of pollen viability (Heslop-Harrison, Heslop-Harrison and Shivanna 1984). It tests principally the integrity of the vegetative cell plasma membrane, and relies on the presence of a non-specific esterase in the pollen cytoplasm. Membranes are permeable to the non-fluorescent, polar molecule, fluorescein diacetate, so it can enter the pollen grain. Active esterases within viable grains cleave the acetate residues, leaving the fluorescent molecule, fluorescein, that accumulates in the grain if the membrane is intact. Hence, only grains with intact membranes and active esterases fluoresce. The test is extremely sensitive, reliable, simple to use, takes only a few minutes to perform, and many papers indicate the results strongly correlate with other pollen tests, including germination and seed set. Some false negative results occur, particularly for pollens that have an ineffective, porous plasma membrane prior to hydration. Therefore it is also important that the pollen is correctly pre-hydrated, since the treatment does not reproduce the slow hydration that occurs on a stigma. The development of the fluorescence may take 30 min or more, but rapid scoring is sometimes needed since the fluorescein may leak from the pollen.

In use, the pollen sample is dispersed on a slide in fluorescein diacetate in a sucrose medium (2 mg/ml fluorescein diacetate in acetone; add dropwise to 2 ml 10-20% sucrose with 1-3 mM H_3BO_3 and $Ca(NO_3)_2$, adjusted to minimize plasmolysis and bursting, until the solution is just persistently cloudy). Under an epifluorescence microscope, 5-30 minutes after staining, viable grains fluoresce a bright yellow-green colour under UV illumination, while inviable grains are only weakly stained. It is easy to score the percentage of viable grains, although is some species the test may not give clear differentiation.

Acetocarmine and DAPI

These stain the pollen grain nuclei, and acetocarmine weakly stains the cytoplasm, and gives good contrast between the grain and surrounding medium. Any nuclear aberrations are clearly visible. The methods are also convenient for examination of archesporial cell development from premeiotic stages to mature pollen.

Fresh pollen (or anther loculi or germinated grains) is dispersed on a microscope slide. For acetocarmine staining, they are suspended in a drop of acetocarmine (1 g refluxed in 45% acetic acid for 24 h and filtered), heated over a sprit flame, squashed and examined after 5-30 min. For DAPI staining, fresh or fixed pollen is stained on a slide in a drop of 0.005% DAPI (diamidinophenylindole), and the nuclei observed under a fluorescent microscope after 5-30 min (UV excitation, blue DAPI fluorescence). Permanent and semi-permanent slides can be made.

Other Staining Methods

Stanley and Linskens (1974) comprehensively review the other staining methods such as iodine-potassium iodide and tetrazolium red. While still useful for particular purposes, the FCR test has largely replaced the other tests because of its accuracy and ease of scoring.

Germination Tests

If pollen is able to germinate and produce a pollen tube, there is a high chance that it is viable and able to fertilize. Hence, for pollens where germination is possible *in vitro,* such tests are valuable.

In vivo *and semi*-vivo *tests*

These tests involve germinating the pollen on an unpollinated stigma. The stigma may be attached to the plant, or excised and planted in a 1% agar medium with 10-20% sucrose. The pollen tubes are stained in the style and observed within it or observed coming from the bottom of the style into the medium. The most widely used stain for pollen tubes is decolorized aniline blue (DAB; 0.1% aniline blue stirred for 24 h in 0.1 M K_3PO_4; see Dumas and Knox 1983). The style is hydrolyzed in strong sodium hydroxide until softened, and mounted on a slide with gentle squashing, under a cover slip in DAB and sometimes 20% glycerin. After 30 min, epi-fluorescent light microscopy with UV or blue excitation will show the pollen tubes by white or yellow fluorescence.

Unlike other tests, hydration on the stigma occurs at the correct rate for the species, so incorrect pre-conditioning is less of a problem (see chapter on Pollen Hydration). Incompatibility reactions will normally be detected by microscopy, but even when they occur still give evidence that the pollen was potentially viable.

In vitro *germination tests*
In vitro germination of pollen is widely used for viability tests, under the generally correct assumption that pollen that germinates and produces a tube *in vitro* is likely to do so *in vivo*, and to fertilize the egg. Pollen grains from many species will germinate and grow on both solid and liquid artificial media. The most widely used media are modifications of the minimal medium of Brewbaker and Kwack (1964), consisting of 1 to 4 mM H_3BO_3 and 1 to 4 mM $Ca(NO_3)_2$ in 0.30 to 0.90 M (10-30% sucrose). Boron and calcium are often the only essential elements, although various workers add magnesium, phosphate buffers, flavenols, non-specific substances such as vegetable juice or yeast extract. More inert ingredients than sucrose (e.g. polyethylene glycol) have been used to prevent the grains bursting. The pollen is dispersed in droplets of liquid media on slides, which are often placed in humid atmospheres hanging upside down (hanging drops) to prevent anoxia, or in 0.5 ml of medium in a tube on a rotator. Alternatively, a semi-solid medium, with the addition of 0.5-1% agar can be used, and the pollen sprinkled on the surface. After 15 min (e.g. rye) to 12 h (e.g. lily), the pollen and tubes are examined by light microscopy. Fixation (in glutaraldehyde, acetic acid : ethanol, or ethanol) will preserve sample for later observation. Germination percentages, growth rate measurements and tube lengths are easily measured.

In general, binucleate pollen germinates readily in such media. Many trinucleate species also germinate, but others - wheat, for example - are recalcitrant and reliable media are not known.

Nevertheless, indications that flavenols such as quercetin can be added to media may give reasonable germination percentages in cultured tobacco pollen (Herbele-Bors et al. 1991) and wheat (Heslop-Harrison and Bardsley, unpublished). New media, e.g. for pistachio pollen germination (Golan-Goldhirsh et al. 1991), are regularly being found.

Pollen Quality

All the methods of testing discussed above depend on many variables and none is able to confirm that a particular sample of pollen is inviable and will not be able to fertilize any plant: they only give a likelihood estimate. Most are unsuitable for field use, and the variability between different samples for the same species, even although fertile, can be high. Nevertheless, cytological methods for testing pollen quality have proved valuable for research and in agriculture, and give unique information about plant fertility.

References

Brewbaker and Kwack (1964)

Cross JW, Ladyman JAR (1991) Chemical agents that inhibit pollen development: tools for research. Sex Plant Reprod 4:235-243.

Lu J, Mayer A, Pickersgill B (1990) Stigma morphology and pollination in *Arachis* L. (Leguminosea). Ann Bot 66: 73-82

Knox RB, Williams EG, Dumas C (1986). Pollen, pistil and reproductive function in crop plants. Plant Breeding Rev 4:9-79.

Dumas C, Knox RB (1983) Callose and determination of pistil viability and incompatibility. Theor Appl Genet 67: 1-10.

Golan-Goldhirsh A, Schmidhalter U, Mueller M, Oretli JJ (1991) Germination of *Pistacia vera* L. pollen in liquid medium. Sex Plant Reprod 4: 182-187.

Herbele-Bors et al. (1991) Como Meeting.

Heslop-Harrison J, Heslop-Harrison Y, Shivanna KR (1984) The evaluation of pollen quality, and a further appraisal of the fluorochromatic (FCR) test procedure. Theor Appl Genet 67: 367-375.

Heslop-Harrison JS, Heslop-Harrison J, Heslop-Harrison Y, Reger BJ (1985) The distribution of calcium in the grass pollen tube. Proc Roy Soc Lond B, 225: 315-327.

Shivanna KR, Linskens HF, Cresti M (1991) Responses of tobacco pollen to high humidity and heat stress: viability and germinability in vitro and in vivo. Sex Plant Reprod 4: 104-109.

Stanley RG, Linskens HF (1974) Pollen: Biology, Biochemistry and Management. Springer Berlin.

MEGASPOROGENESIS AND MEGAGAMETOGENESIS

M.T.M.Willemse
Dept. Plant cytology and morphology, Agricultural University,
Arboretumlaan 4, 6703 BD Wageningen, The Netherlands.

Introduction.

Megasporogenesis and megagametogenesis lead to the formation of the embryo sac. It includes the development of a highly differentiated plantlet in function of the sexual reproduction of angiosperms. After a meiotic division and cell isolation the development from the unicellular stage to the multicellular embryo sac represents the life cycle of the haploid plant. This life cycle from megaspore to megagametophyte with the megagamete takes place in and is strongly related to the mother-plant, the sporophyte.

The life cycle of the megagametophyte.

The functions of the megagametophyte are:

-to maintain contact with the mother plant to get nutrients

-to prepare the acceptation of the pollen tube

-to realize the double fertilization

-to prepare the start for the formation of the embryo and endosperm.

This implies that the megagametophyte is a very specialized organism!

The way to reach this functions occurs gradually during the life cycle in different stages in cooperation with the mother-plant. Variations on the different steps and sometimes phases leads to different types of development and embryo sacs. The most common is the Polygonum type.

During the life cycle the following stages with their main phases can be distinguished:

I. Preparative during megasporogenesis: the formation of one haploid isolated cell.

planning stage: with the formation of the archespore cell and induction of the meiosis.

transition stage: with the meiosis, transition to haploidy and to cell polarity.

selection stage: with the isolation and selection of the type of functional megaspore and uptake of storage for further development.

II. Differentiation during megagametogenesis: the formation a pollen tube acceptation apparatus, the preparation of the double fertilization and endosperm.

maturation stage: with nuclear divisions and the transition from coenocytic to cellular megagametophyte and cellular differentiation to a functional but not activated embryo sac

fertilization stage: with the activation, the acceptance of the pollen tube, the fusion of gametes and the transition to diploidy.

The surrounding tissues of the mother-plant, the nucellus and integuments also develop during megasporo- and -gametogenesis.

-During the megasporogenesis the nucellus and the inner integument are formed but not yet completed. In the funiculus the vascular strand differentiates. In some plants the development of the epistasis or hypostasis can start.

-During megagametogenesis the second integument is formed and covers the first stretching integument. After the formation of the megaspore a part of the nucellus is still in development but in the micropylar region locally some nucellar cells may degenerate to give space to the developing megagametophyte.

Steps during the different stages and phases during megasporogenesis and megagametogenesis.

Megasporogenesis.
During the planning stage two phases are distinguished: the phase of archespore cell formation and the induction of the meiosis.
Archespore cell formation:
1. After a division as first step the subepidermal nucellar cell starts to increase the cytoplasmic content. The following enlargement of this archespore cell is accompanied by the increase of the plasma and its organelles, as the number of ribosomes. In the cell centre the nucleus becomes more voluminous.
Induction of meiosis:
1. The enlarged archesporial nucleus gets the chromosomal leptotene configuration at the onset of meiosis. The factors leading to meiosis are still unknown.
2. At zygotene in the nucleus vacuole-like structures become visible and the perinuclear membrane can show enlargements, a possible sign for a renewed interaction between nucleus and cytoplasm.
3. During induction, the microtubular system in the cell is at random organized. It seems that mainly the chalazal part of the archespore enlarges

while the position of the nucleus remains fixed. In this way a cell polarity is build up by unequal cell stretching.

During the transition stage two phases can be distinguished: the completion of meiosis and the transition from diploid to haploid status.

Transition from diploid to haploid status:

1. During zygotene/pachytene commonly around megameiocytes have a callose wall which appears at the micropylar side and extends in chalazal direction. In this way an apoplastic local barrier to high molecular products is formed and a partial cell isolation is realized. In most cases the callose wall remains incomplete and at the chalazal side plasmodesmata can be still intact and functional.

2. The megameiocyte is enlarged and the plastids and sometimes the mitochondria are localized near the cell poles with a higher number in the chalazal region. After the first meiotic division the existing unequal organelle distribution becomes more expressed by the nuclear repulsion and more cytoplasm is positioned near the chalazal part. During this repositioning of the organelles the microtubules are not directly involved.

3. The number of ribosomes decreases in most types of megameiocytes, indicating the removal of diploid gene information in the cytoplasm.

Completion of meiosis:

1. After the second meiotic division cellularization results in a linear tetrad with four megaspores, all separated from each other by callose walls. In those cases where such cellularization does not occur a four nucleated megaspore is formed.

During the selection stage two phases are present: the formation of a functional megaspore and storage production.

The formation of the megaspore.

1. The functional megaspore becomes selected. In those cases where four megaspores are formed, the following factors are involved in this selection:

-Genetical: In *Oenothera* it is proved that the functional megaspore is determined by the genetical background present.

-Hormonal: a hormonal influence is supposed to be involved in this selection.

-Polarity: the prepared cell polarity results in an unequal cytoplasm and organelle distribution.

-Nutrients: the way of nutrient supply and the presence of an hypostasis may interfere with the position of the selected megaspore. Especially the presence of a callose wall can influence a selective transport.

2. Using heatshock treatments in anthers of *Hyacinthus* sp. of *Tulipa* sp. haploid embryo sacs can be formed. This indicates that the moment of sex

determination can occur from completion of the meiosis.

Storage production.

1. As result of the selection, the functional megaspore gets starch and lipids as storage products and the number of ribosomes increases. It is suggested that the functional megaspore uses the breakdown products of the other three megaspores after their degeneration, inclusive the callose wall. This storage will be used in the following stages.

Megagametogenesis.

During the <u>maturation</u> stage three phases can be distinguished: the first is the formation of the coenocyte with the nuclear divisions, the second is the cellularisation of the coenocyte and the last phase is the differentiation of the individual cells.

The formation of the coenocyte.

1. The functional megaspore enlarges by vacuolation. The vacuoles surround the central nucleus and fuse. After the first mitosis and the following mitoses the vacuole is in the centre of the coenocyte and will enlarge.

2. The central nucleus divides and the two daughter nuclei become positioned each at a pole of the coenocyte. Dependent on the type of development, more nuclear divisions will follow. The positioning of these nuclei is important and probably determined by the filamentous cytoskeleton. Microtubules are involved in the division of the nuclei.

In the *Polygonum* type three nuclei will be positioned at the micropylar pole and one near these nuclei. At the chalazal pole also three nuclei and one near to these nuclei are positioned.

3. During the formation of the coenocyte, the storage products will disappear. Probably there is no increase in the number of cell organelles.

4. During the nuclear divisions the change in polarity in the coenocyte can be realized by nuclear fusion or degeneration in the chalazal part.

The cellularisation of the coenocyte.

1. In the *Polygonum* type cell walls will be formed between the three nuclei at both the micropylar and chalazal pole. This cellularization results in three micropylar cells, which form the future egg apparatus and at the chalazal pole in the three antipodal cells. Two nuclei remain present in the central cell, the remnant of the coenocyte.

The differentiation to the embryo sac.

1. The egg apparatus consists of two synergids and one egg cell. All these cells make contact with the micropylar part of the embryo sac.

-In the egg cell the nucleus is firstly positioned in the upper part of the cell and the vacuole enlarges in the basal part of the cell. Most organelles are located near the nucleus. Before fertilization the nucleus and surrounding cytoplasm move to a chalazal position in the cell. Cell organelles will increase in number before fertilization and in the zygote. The egg cell functions as the female gamete but lacks storage as female gametes possess commonly.

-The two synergids will produce the filiform apparatus. Each cell get develops a large lateral transfer-like cell wall which thickens mainly in the upper part of the cell. The endoplasmic reticulum as well as dictyosome vesicles are involved in the formation of this wall. The composition of this wall is heterogenous and consists mainly of matrix material and few cellulose fibrils. The microtubules are abundant in these cells to realize and maintain their shape. The cell walls around the synergids as well as the egg cell are incomplete. At the border with the upper part of the central cell the cell walls are very thin or absent.

The function of the synergids is to receive the pollen tube. Commonly one of the synergids will degenerate. The production of micropylar exudate including a pollen tube attractant from the synergids as well as nucellar or inner integument cells cannot be excluded.

2. The central cell is coenocytic but the two nuclei will fuse to one diploid nucleus which commonly is positioned near the egg apparatus. Around the nucleus the cytoplasm gets amylum as storage product. The central cell will fuse with one of the sperm cells.

3. The antipodal cells show a diversity in number and composition. Usually the antipodal cells contain abundant cytoplasm and have only small vacuoles. The cytoplasm is rich in ribosomes. In some antipodals the first sign of degeneration can be observed by the hyperplasia of plastids and mitochondria. Other can store storage products as amylum, lipid and proteins, which are used after fertilization. The antipodals function mainly in the transfer of nutrition via the symplast to the embryo sac. In the cell wall bordering the central cell plasmodesmata are present.

During the fertilization stage there is a phase of activation of the embryo sac, the acceptance of the pollen tube and the fusion of gametes and the transition to diploidy takes place.

Activation of the embryo sac.

1. The pollination stimulus evokes the activation of the pistil, including the embryo sac.

Signals are supposed to function in the activation. In the embryo sac this activation is expressed by the movement of the nuclei of the central cell and egg cell and in the onset of protein synthesis.

Acceptance of the pollen tube.

1. The prepared micropylar and nucellar pathway lead the pollen tube to the degenerated synergid. The tube passes between the filiform apparatus and will burst inside the synergid.

Fusion of the gametes.

1. With the fusion of one male gamete with the egg cell and the other with the central cell, the megagametophyte converts. In the zygote the cytoplasmic transition from haploid to diploid occurs, the polyploid central cell, the primary endosperm cell has its own genome. Only the antipodals remain haploid and if their number is low, they will degenerate.

Regulation of megasporogenesis and megagametogenesis.

In embryo sac formation one main line of regulation are the subsequent divisions: the meiosis and mitosis. In case of the formation of a haploid embryo sac a meiosis should take place. Thereafter in case the cellularisation remains absent, the number of four nuclei can be sufficient to make a functional embryo sac. In case of a binucleate of one nucleate megaspore at least one mitosis should be expected.

Commonly the number of three mitoses leads to a seven celled embryo sac, but more mitoses can occur, which result in the formation of a higher number of antipodal cells.

In the regulation of the nuclear divisions there is at first meiosis and secondly a variation of one to more mitoses.

The other distinctive regulatory steps are expressed during the progress of nuclear divisions and can be related to the different stages.

Megasporogenesis.

In the planning stage variation of the position and number of archesporial cells can result in twin embryo sacs. But in this stage no further events influence the regulation to the formation of a embryo sac.

In the transition stage the position of the nucleus towards the micropyle is the expression of the chalazal elongation of the cell, leading to an unequal distribution of the cytoplasm.

This regulates the ultimate cytoplasmic contents of the various megaspores.

In the selection stage the first determining regulation is the absence or presence of cell wall formation.

-If no cellularization occurs, the megaspore will contain four nuclei. This results in the tetrasporic type as occurs in the *Adoxa, Fritillaria, Plumbagella, Drusa, Penaea, Piperomia* and *Plumbago* type.

-If only during the meiosis I a cell wall is formed, the bisporic type as in *Allium* and *Endymion* originates.

-If after the meiosis II a linear tetrad is formed by complete cellularization the *Oenothera* and *Polygonum* types arise.

-Exceptional is a change in the orientation of the cell walls after meiosis II leading to a wall or walls perpendicular to the existing cell wall after meiosis I. This event leads to abberations in the developmental pattern but does not influence the selection of a functional megaspore.

The position of the wall divides the cell in two equal parts or is already in favour of the position of the functional megaspore as in *Oenothera* and *Podostemon*.

The position of the nuclei as result of the type of nuclear repulsion during the meiosis II, and directed by the position of the spindle is commonly linear. Except in tetrasporic types as *Adoxa, Fritillaria* and *Plumbagella*, where a quadrant of nuclei is present and in *Penaea, Piperomia* and *Plumbago* with a more diamond position.

The pattern of the megaspore degeneration, at the micropylar side for the *Polygonum, Allium* and *Podostemon* types and at the chalazal part for the *Oenothera* type, is an effect of regulation which leads to a preferential position of the functional megaspore.

Megagametogenesis.

The maturation stage is very susceptible for alternative or special regulations. During the coenocytic phase the regulation of nuclear position is determined by the mitoses and expressed in the position of the spindle. Different spindle positions can be observed and mark the different types. The number of nuclear divisions is commonly determined, although in *Chrysanthemum* a variation in the level of polyploidy of the central cell nucleus and in the number of antipodal cell is found.

Directed nuclear movement independent of nuclear divisions takes also place. The mechanism of these movements may be the fibrillar cytoskeleton system. Comparable with the spindle positions, this movement organizes the ultimate nuclear position and is therefore discriminating for the different types.

Involved in the positioning of the nucleus is also the process of vacuolation and ultimate position of the vacuole. In most types the vacuole gets a central position. But in early coenocytic stages more vacuoles permit a more complex localization of the cytoplasm and nuclei.

All these systems cooperate in the ultimate position of the nuclei at the micropylar or chalazal pole. With this complex system the position of nuclei can be shifted easily. Such abberations or alternatives are frequently found as in *Tamarix* and *Sambucus*.

Nevertheless, it will result in a functional megagametophyte.

During the coenocytic phase nuclear degeneration takes place in the chalazal part of the cell in the *Podostemon* type.

Nuclear fusion of the chalazal nuclei in the *Fritillaria* and *Plumbagella* type are proceeded by directed nuclear migration to the chalazal end.

Next to the position of the nuclei, the composition of the cytoplasm is a determining factor for the cellularization and following differentiation.

Regulation of the megasporo- and megagametogenesis is till now unknown, but a relation with the environmental conditions can be expected.

Nutritive pathway during megasporogenesis and mega-gametogenesis.

The knowledge about the composition and the pathway of the nutrients in the developing ovule, is very limited.

It has been suggested that during the selection of the megaspores, resulting in a breakdown of the callose wall and the degeneration of the cells, the breakdown products are re-used by the developing young functional megaspore.

The nucellus originates from and develops in close contact with the placental tissue. Via symplastic and apoplastic transport the growing cells get their nutrition. A local apoplastic barrier is the hypostasis if phenol containing substances as lignin or suberin are present in the cell wall. The hypostasis appears during the end of the meiosis. From that period the basal nutrient flow is partly changed in a symplastic flow only. If the plasmodesmata between the megaspore become blocked, the symplastic transport stops.

A selection in the pathway of nutrition at the chalazal end of the embryo sac can be the result. During the coenocytic phase the cell increases in volume by the uptake of water.

If this is supplied lateral, the hypostasis prevents also a drop of water content of the coenocyte.

Considering the position of glucose and amylum storage, a banding pattern in the nucellus is build up when the embryo sac develops in *Gasteria* sp. This means a preferential place in the nucellar tissue. The bands of glucose are positioned at the chalazal part surrounding the hypostasis and near the basal part of the micropyle. Starch is present in the upper part of the nucellus and in the central cell. These patterns suggest a directed transport to a concentration point in the top of the nucellus and for glucose at the basal part of the nucellus also.

Transfer walls can be present in the top of the embryo sac and along the central cell.

Especially after fertilization the nutrition of the embryo via the suspensor takes place via the top region of the nucellus. From a structural and functional point of view the transport has a main direction toward the top.

Exceptional is the use of the nutrients outside the nucellus from the placental locule, which runs either via the pollen tube or through haustoria formed by the synergids or other cells of the embryo sac.

From labelling experiments with fluorochromes and radioactive amino acids also a lateral transport to the embryo sac is stated.

The vascular strand develops during the formation of the funiculus and ends commonly at the chalazal part of the nucellus.

If a row of ovules is present on the placenta, a gradient in development can be expected.

There are indications that at first the ovules near the base are mature.

References

Maheshwari P Ed. Recent advances in embryology of angiosperms Int Society of plant morphologists, University of Delhi, 1963

Rutishauser A Embryologie und Fortpflanzungsbiologie der Angiospermen. Springer-Verlag, Wien New York, 1969

Batygina T et al. Eds. Comparative embryology of flowering plants. Leningrad auka Leningrad Branch, Winteraceae-Juglandaceae 1981, Phytolaccaceae-Thymelaeceae 1983, Brunelliaceae-Tremandraeae 1985, Davidiaceae-Asteraceae 1987, Butomaceae-Lemnaceae 1990

Johri BM Ed. Embryology of angiosperms. Springer-Verlag, Berlin Heidelberg New York Tokyo. 1984

Sedgley M, AR Griffin AR Sexual reproduction of tree crops. Academic press Harcourt Brace Jovanovich Publishers, London San Diego New York Berkely Boston Sydney Tokyo Toronto 1989

The Angiosperm Stigma

J.S. Heslop-Harrison
Karyobiology Group
Department of Cell Biology
JI Centre for Plant Science
Norwich NR4 7UJ England

Introduction

The angiosperm stigma is an efficient structure with both morphological and physiological adaptations that enable pollen capture, hydration and germination. The stigma surface may play a vital part in controlling interspecific hybridization and in regulating compatibility relationships within species. The structural and physiological features of the pollen capturing surfaces vary considerably between families and sometimes within families. This chapter aims to show the range of morphological variation in stigma types found throughout the angiosperms, to enable classification of stigmata into the major recognized groups, and to indicate the relationship between structure and function.

The Flower

Floral structures are among the most important and variable characters used for plant identification. Indeed, many species can only be separated on floral characteristics. The variability arises because the flower is the structure with the highest complexity of any plant part - in number of sub-structures, tissue organization, and number of genes expressed. Despite the importance of the flower in taxonomy, little attention has been paid to one of its major sub-structures: the stigma. This contrasts with the pollen, where several textbooks exist with pictures and keys for plant identification based on pollen alone. This does, though, reflect current use of the character: pollen is often preserved in dry or wet conditions, potentially for thousands of years, and is valuable for the identification of species composition of ancient ecosystems, and hence for discovery of the climate and other variables.

The Stigma

The stigma is the receptive surface of the style that collects the pollen and enables its hydration and germination. The style connects the stigma to the ovule and includes the pollen tube transmitting tract. The style is involved in pollen tube guidance, nutrition and incompatibility responses. Angiosperm stigmata (or stigmas) are structurally very diverse and the surfaces adapted for pollen grain capture differ widely both in the morphology of the receptive cells and in the amounts of surface secretion. The first large scale systematic survey of stigmata from 250 families was carried out by Heslop-Harrison and Shivanna (1977), and a major reassessment of the taxonomic work on stigmata was presented by Heslop-Harrison (1981). These works, to which considerable reference is made here, described the major surface characteristics and morphologies of stigmata from 1000 species belonging to 900 genera, and developed a classification scheme for the various types of stigma (Table 1). The system divides species into two major groups: 1) those with stigmata that have a wet surface, bearing a fluid secretion, and 2) those where the stigma surface is dry and lacks any surface secretion; subdivisions separate species based on the presence of trichomes or papillae.

Table 2 gives the stigma types of species in a few major families that are of economic, scientific or horticultural interest. A much fuller listing of the characteristics of 250 families is given by Heslop-Harrison and Shivanna (1977).

Study of the Stigma

The stigma is one of the most short-lived structures of a plant. Each stigma remains receptive to pollen for a few days at most, and in some species may be functional for only a few minutes after pollination. Many characteristics of the stigma can only be studied when it is in the receptive, mature state. Particularly in species with wet stigmata, substantial maturation may occur within the hours, or even minutes, immediately before the stigma becomes receptive. Surface fluids may be rapidly secreted, but in some species the stigma begins to autolyse

Table 1. A classification of Angiosperm stigma types based on the amount of secretion present during the receptive period and the morphology of the receptive surface. Abbreviations for each major type are given in the right hand column. (After Heslop-Harrison 1981).

Surface Dry
Receptive cells (trichomes) dispersed
  on a plumose stigma.................................. D Pl
Receptive cells concentrated in
  ridges, zones or heads
    Surface non-papillate .... D N
    Surface distinctly papillate
      Papillae unicellular .... D P U
      Papillae multicellular
        Papillae uniseriate .... D P M Us
        Papillae multiseriate.................. D P M Ms

Surface Wet
Receptive cells papillate;
  secretion moderate to slight,
  flooding interstices .... W P
Receptive cells non-papillate;
  secretion usually copious .... W N

as soon as it becomes receptive. Therefore many characters can only be studied from fresh, mature stigmata, and the study of fixed or dried stigmata is inadequate. In agriculture, the effective pollination period (EPP) is an important character, and it is measured to find the time available for pollination of a plant variety. In some species, there may be little visible difference between receptive and non-receptive stigmata. For example, in avocado, the stigma accumulates callose as it senescences and becomes non-receptive (Sedgely 1977), although the difference between a receptive and non-receptive stigma is not visible without staining. Testing for stigma receptivity is difficult, although some methods are available (reviewed by Knox, Williams and Dumas 1986).

Observations of stigma surface morphology and secretions can often be made with a dissecting microscope or hand lens. However, the scanning electron microscope (SEM) is extremely valuable for observing and recording stigma morphology because of its high resolution, large depth of field and ability to examine untreated stigmata. The most informative SEM images are obtained using fresh, unfixed, and uncoated material; procedures including fixation, drying and coating are usually unnecessary and generally add artifacts. A fresh stigma is excised and planted on a specimen stub in a viscous conductive solution - fish or animal derived glues with high salt concentrations, or graphite particle suspensions are usually suitable. The fresh specimen must be observed and recorded quickly once placed in the microscope chamber to avoid dehydration and charging.

For cytochemical and ultrastructural studies of the stigma, a wide range of techniques is used. Squashed preparations, often made as thick mounts, are valuable to show cellular morphology and differentiation in the stigma - unicellular or multicellular papillae can be observed easily, and the pollen tube pathways examined. Cytochemical studies (Pearse 1972; Gurr 1965) enable ions, proteins, lipids and carbohydrates to be stained specifically, and localized in preparations from fresh, unfixed material. For higher resolution studies, fresh, fixed or frozen stigmata can be studied by electron microscopy techniques (see Shivanna, Ciampolini and Cresti 1989; Heslop-Harrison 1990).

The Dry Stigma (D)

A few genera, such as *Euphorbia,* have smooth dry domes of cells to which the pollen grains adhere and directly hydrate (DN), but most dry stigmata have trichomes (figure 1), unicellular (figure 2) or multicellular papillae. Pollen grains have a size appropriate to the papilla size and

Table 2. Representative families that include genera with the stigma characteristics shown in table 1. Some families have genera that are in more than one subdivision. See Heslop-Harrison and Shivanna (1977) for a classification of 900 genera.

Gramineae	D Pl
Betulaceae	D N
Crucifereae	D P U
Ranunculaceae	D P U
Malvaceae	D P U
Liliaceae	D P U
Oleaceae	D P M Us
Geraniaceae	D P M Ms
Leguminosae	W P
Ericaceae	W P
Solanaceae	W P
Orchidaceae	W P
Liliaceae	W P
Ericaceae	W N
Umbelliferae	W N

adhere on or between the papillae before hydration and germination (see Chapters on Pollen Hydration and Cresti, Pollen Germination, this volume).

The Dry Plumose Stigma (DPl)

Plumose stigmata are characteristic of the grasses (Gramineae; see figure 1). They are well adapted to collecting wind dispersed pollen, and there is a strong co-adaptation between the size of the pollen grains, the morphology of the stigma and the spacing and positioning of the trichomes. The area of the receptive surface varies from a few mm^2 in small species to over 1000 mm^2 in *Zea mays*, where the entire "silk" (stigma) can capture pollen. All grass stigmata have a dry surface, covered with a protein containing pellicle, and are adapted to enable rapid hydration and germination of the pollen grain: Heslop-Harrison (1987) has shown how pollen of rye may rehydrate and germinate within 2 min of attachment to the stigma.

The Dry Papillate Stigma (DP)

A few workers have examined the ultrastructure and cytochemistry of the stigma of species with dry, unicellular papillate (DPU) stigmata. An early study by Heslop-Harrison and Heslop-Harrison (1975) showed that the stigma papillae of *Crocus* had a loose, chambered cuticle overlaying a thick pectocellulosic wall. Later work (Heslop-Harrison 1987) showed the penetration of the pollen tube under the stigma cuticle in *Crocus*. Shivanna *et al.* (1989) described the thin, outer pellicle of the stigma of *Hypericum calycinum*, and showed that the cell wall of the papillar cells consisted of an outer, loosely woven fibrillar layer and more dense inner layer with compact fibrils. These wall characteristics enable pollen tubes to penetrate the cuticle, often near the base of the papillae, and enter the transmitting tract of the style, which may be either hollow or filled with cells. Dry papillate stigmata are generally (although not exclusively) associated with sporophytic self incompatibility, where the incompatibility factor carried on the surface of the pollen relates to the paternal genotype, and inhibition occurs at or near the stigma surface (see Thompson and de Nettancourt, this volume).

Some species with dry stigma surfaces have a copious secretion under a detached surface cuticle. The cuticle is normally ruptured by pollinating insects, which releases the secretion so that pollen hydration and germination can occur (Lord and Heslop-Harrison 1977). Under field conditions, different lines of *Vicia* show varying levels of autofertility, and analysis shows that the structure of the stigma varies between the lines. Those with thin cuticles over short papillae, that easily rupture, are much more autofertile than lines with thick cuticles and long papillae. Therefore, variations in stigma characters can be important for reproductive isolation of lines, and lead to differences in the field performance of crops.

The Wet Stigma (W)

Wet stigmata have a copious surface secretion. The secretion can be in a crater, or stigmatic cup, as in some Legumes (Owens 1990; figure 3), or on a dome, as in some lilies. Often, the secretion is far from homogeneous, consisting of both hydrophobic and hydrophilic components, and including lipids, proteins and carbohydrates. Konar and Linskens (1966a, b) showed a remarkable adaptation in the wet stigma of *Petunia*: it was stratified with a surface, lipid rich layer and a thin aqueous layer underneath. The pollen lands on the lipid layer and sinks through to the aqueous layer where hydration begins. This adaptation is also found in other Solanaceae, but not Rosaceae. Owens (1989) showed that there were dispersed lipid drops in the surface secretion of some legumes. The secretion also included sloughed off cells that arose from the loose, secretory tissue underneath the secretion; these secretory cells, and the large, intercellular spaces, could be seen clearly after critical point drying. Slater and Calder (1990) examined the ultrastructure of the detached cells in the stigmatic secretion of an orchid (*Dendrobium speciosum*) and found that the cells had all the characteristics of active secretory cells including extensive networks of endoplasmic reticulum, dictyosomes and vesicles.

Since materials carried on the surface of pollen grains are dispersed in the secretion, wet stigmata are associated only with gametophytic self-incompatibility systems, where the pollen genotype itself determines rejection or acceptance, and the rejection reaction usually occurs in the style.

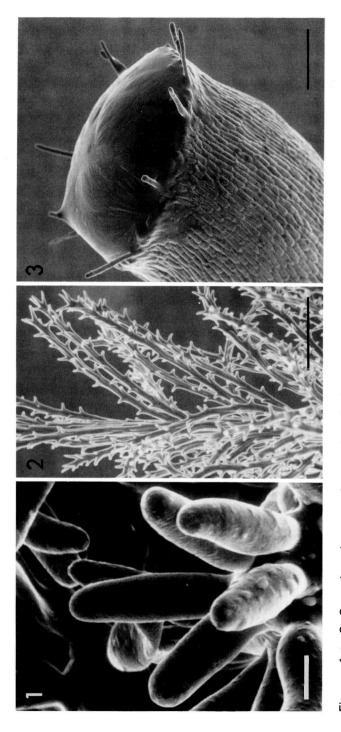

Figures 1 to 3. Scanning electron micrographs of stigmata characteristic of the major groups in table 1. 1. *Crocus chrysanthus* (Iridaceae) with a dry, papillate, unicellular (DPU) stigma; bar 25 μm; (courtesy Dr Y. Heslop-Harrison, Aberystwyth). 2. A dry plumose (DPl) stigma from wheat (*Triticum aestivum*), characteristic of grasses. Most grass stigmata are bifurcated (not shown); bar 200 μm. 3. A wet papillate (WP) stigma from a *Senna nemophila* (Leguminoseae; courtesy Dr S.J. Owens, Kew); bar 100 μm.

Stigma Turgidity

The stigma is usually strong and, particularly in the grasses with plumose stigmata, exerted from the flower. These characteristics are normally given by turgidity, and after pollination, the turgid stigmatic tissues often collapse and desiccate rapidly. While the surfaces of most plants are heavily protected from dehydration and pathogen attack, the stigma cannot be protected by a thick cuticle or wax since it must capture and hydrate pollen, and allow eventual penetration of the pollen tube into the intercellular spaces of the pollen tube transmitting tract. In the grasses, the stigma must remain turgid in any atmosphere and cannot be well protected when pollen is shed. Results from cytochemical and energy dispersive X-ray analysis of the elements present in the stigma show that potassium and chloride ions together account for 60% of the osmolality of stigma sap in *Pennisetum* (Heslop-Harrison 1990; Heslop-Harrison and Reger 1986). Thus these ions are very important for maintaining the turgidity of the stigmatic cells and, in this system as in the anther filament (Heslop-Harrison, Heslop-Harrison and Reger 1987), it is probable that regulation of potassium ion movement provides a rapid and sensitive method to control the cell osmoticum and to expand the stigma. A secondary effect of the high ion concentration makes the stigma electrically conductive, so that it does not charge strongly when examined under the electron beam of the SEM.

Some species in the Boraginaceae have an adaptation that may be related to water conservation. The papillar cells of the stigma have caps that touch to form a "roof". Pollen landing on the roof does not hydrate, while that which is forced between the papillae (perhaps by insects walking on the surface), hydrates and germinates rapidly, presumably in the moist atmosphere under the roof (Heslop-Harrison 1981). Hence the outer surface, forming the roof of each papillar cell, can have a thickened, impermeable wall, while the stalk, underneath the roof, can have a more permeable wall that enables hydration of pollen and penetration of the pollen tube.

Pathogen Protection

The stigma also provides a potential point for fungal spores to germinate and penetrate the plant. In the cereals, ergot is a minor fungal disease that can enter through the stigma, and a fungal mass (sclerotium or ergot) replaces the grain. Susceptibility of the plant seems to be related to the time the flower remains open, so plants with flowers that are pollinated and then close soon after the stigma becomes receptive are rarely infected. Male sterile plants of the normally cleistogamous barley, where the flowers remain open for a long time, are extremely susceptible to ergots - up to 76% of the heads may become infected, although fertile barley lines are rarely infected in the field (Agrios 1988). In some species, including *Zea mays,* the stigma abscises after pollination (Heslop-Harrison, Heslop-Harrison and Reger 1985), while in other species it becomes flaccid or dehydrates. The reasons for the generally high level of resistance to stigma infection are unknown, but it seems likely that there are several barriers to infection used by plants.

Stigma Characteristics

Although characteristics of the stigma have rarely been considered in taxonomic and phylogenetic studies, clearly there are many regularities in stigma classification within genera and families. Although some families have stigmata of more than one type, evolutionary trends and intermediate states are often evident so stigma morphology can be used, like pollen morphology, for classification studies. Stigmata show substantial structural adaptations, and co-adaptations with pollen, to ensure efficient capture of pollen from the same species and rejection of alien, and sometimes self, pollen. Further structural features are important to enable pollen hydration, germination and pollen tube penetration, while minimizing water loss and pathogen invasion. Understanding the form and function of the stigma and its interaction with the pollen is a vital part of any study of sexual plant reproduction.

References

Agrios GN (1988) Plant Pathology. 3rd ed Academic Press New York.
Gurr E (1965) The Rational Use of Dyes in Biology. Williams and Watkins Baltimore.
Heslop-Harrison J (1987) Pollen germination and pollen-tube growth. Int Rev Cytol 107: 1-78.
Heslop-Harrison J, Heslop-Harrison Y (1975) Fine structure of the stigmatic papilla of *Crocus*. Micron 6:45-52.

Heslop-Harrison J, Heslop-Harrison Y (1983) Pollen-stigma interaction in the Leguminosae: The organization of the stigma in *Trifolium pratense* L. Ann Bot 51:571-583.

Heslop-Harrison JS (1990) Energy dispersive X-ray analysis. pp 244-277 in Modern Methods in Plant Analysis New Series, Volume 11, Ed HF Linskens and JF Jackson Springer Berlin.

Heslop-Harrison JS, Heslop-Harrison Y, Reger BJ (1987) Anther-filament extension in *Lilium*: Potassium ion movement and some anatomical features. Ann Bot 59:505-515.

Heslop-Harrison JS, Reger BJ (1986) Chloride and potassium ions and turgidity in the grass stigma. J Plant Phys 124:55-60.

Heslop-Harrison Y (1976) Localisation of concanavilin A binding sites on the stigma surface of a grass species. Micron 7:33-36.

Heslop-Harrison Y (1981) Stigma characteristics and angiosperm taxonomy. Nordic J Bot 1:401-420.

Heslop-Harrison Y (1990) Stigma form and surface in relation to self-incompatibility in the Onagraceae. Nordic J Bot 10:1-19.

Heslop-Harrison Y, Heslop-Harrison J, Reger BJ (1985) The pollen-stigma interaction in the grasses. 7. Pollen-tube guidance and the regulation of tube number in *Zea mays* L. Acta Bot Neerl 34:193-211.

Heslop-Harrison Y, Shivanna KR (1977) The receptive surface of the angiosperm stigma. Ann Bot 41:1233-1258.

Knox RB, Williams EG, Dumas C (1986) Pollen, pistil and reproductive function in crop plants. Plant Breeding Reviews 4:9-79.

Konar RN, Linskens HF (1966a) The morphology and anatomy of the stigma of *Petunia hybrida.* Planta 71:356-371

Konar RN, Linskens HF (1966b) Physiology and biochemistry of the stigmatic fluid of *Petunia hybrida.* Planta 71:372-387

Lord EM, Heslop-Harrison Y (1984) Pollen-stigma interaction in the Leguminosae stigma organization and the breeding system in *Vicia faba* L. Ann Bot 54:827-836.

Owens SJ (1989) Stigma, style, pollen and the pollen-stigma interaction in *Caesalpinioideae.* Advances in Legume Biology Monogr Syst Bot Missouri Bot Gard 29:113-126.

Owens SJ (1990) The morphology of the wet, non-papillate (WN) stigma form in the tribe *Caesalpinieae (Caesalpinioideae: Leguminosae).* Bot J Linn Soc 104:293-302.

Owens SJ, Lewis GP (1989) Taxonomic and functional implications of stigma morphology in species of *Cassia, Chamaecrista* and *Senna (Leguminosae: Caesalpinioideae).* Plant Syst Evol 163:93-105.

Pearse AGE (1972) Histochemistry - Theoretical and Applied. 2 vols, 3rd ed Churchill, London.

Sedgely M (1977) Reduced pollen tube growth and the presence of callose in the pistil of the male floral stage of the avocado. Sci Hort 7:27-36.

Shivanna KR, Ciampolini F, Cresti M (1989) The structure and cytochemistry of the pistil of *Hypericum calycinum*: The stigma. Ann Bot 63:613-620.

Slater AT, Calder DM (1990) Fine structure of the wet, detached stigma of the orchid *Dendrobium speciosum* SM. Sex Plant Reprod 3:61-69.

TRANSPORT MECHANISMS OF POLLEN - A SHORT REVIEW

E. Pacini
Dipartimento di Biologia Ambientale
Università di Siena
Via P.A. Mattioli 4
53100 SIENA
Italy

Definition

Pollination is a process typical of gymnosperms and angiosperms. Gymnosperm pollen grains are produced by the pollen sacs and land on the micropyle of the ovule, very close to the existing or future female gametophyte. Angiosperm pollen is produced by anthers and lands on the stigma, even as far as several centimeters from the female gametophyte contained in the ovary.

Prelude

The process leading to anther dehiscence is the result of a programme involving all the different types of cells in the anther; some of these processes start very early in anther-pollen development (Bonner & Dickinson 1989). During the last phases of pollen development, the water content of the whole anther decreases. Before dehydration is complete, pollenkitt, the product of tapetum degeneration, is deposited on the pollen grain surface (Keijzer 1987). Tryphine is deposited on the pollen grain surface earlier, at the microspore stage (Pacini 1990a, b). Tryphine and pollenkitt serve several purposes, two of which are to cause clumping of the pollen grains and to facilitate their adhesion to pollinators. As a consequence of anther dehydration pollen grains become dormant, but this dormancy is not comparable to that of seeds, lasting only of the order of minutes or days.

Flower and anther anthesis

Pollination occurs when the anthers open to expose the pollen, and the stigma is receptive. Two antheses occur during blooming: that of the flower and that of the anthers (Keijzer 1987). These two processes are not always contemporaneous, and hermaphroditic flowers are often dichogamous, i.e. the parts of either sex are not receptive at the same time (fig. 1). They are often proterandrous because they ripen centripetally, but many sequences of male and female receptivities are known in angiosperms (fig. 1). Flowers disposed in inflorescences may bloom contemporaneously or otherwise. When the flowers of an inflorescence blooms all together and both sexes are receptive, self pollination is possible, unless there are devices to prevent it. When the flowers of an inflorescence do not bloom together and the sexes are not receptive at the same time, self pollination is unlikely. In such cases the inflorescence is visited by pollinators. Irrespective of the pollen vector, the flowers and anthers open when there is the highest probability of pollination. In some entomophilous species, the flowers open when pollinators are about.

Duration and pattern of floral anthesis
The duration of flower anthesis varies according to pollination syndrome. It is usually very short in anemophilous species like grasses (Dowding 1987); it is very long in species like orchids having a typical and specialized entomophilous syndrome, lasting as long as 60 days in *Lemboglossum maculatum* (Clifford & Owens 1988). This means that pollen grains maintain their viability for the same amount of time. Anthesis is continuous in the case of orchids. In species such as the broad bean (*Vicia faba*), there are seven opening and closing rhythms of the flower (Perryman & Marcellos 1988). A very short period of receptivity in an entomophilous species occurs in *Cucurbita pepo*, the pollen of which is exposed only from 6 AM to noon, when the corolla wilts; pollen viability decreases sharply during anthesis (Nepi & Pacini, unpublished).

Pollen vectors
Pollen vectors, their specificity, efficiency and the average distances reached by pollen grains are reported in Tab 1. Anemophilous and entomophilous species have ecological, morphological and cytological peculiarities (Tabs 2-5). Generally speaking, anemophily is more expensive for the species because enough pollen must be produced to enable some to reach the right stigma. On the other hand, entomophilous species must produce substances or devices able to attract pollinators, and nectaries sink a lot of energy (Pyke 1991). Dowding (1987) reviewed and listed the advantages and disadvantages of wind pollination. Pollination in gymnosperms is anemophilous and a specificity between the shape of the cone and that of the pollen has been noticed (Niklas 1985). When a cloud of pollen grains passes a female cone, it is splits up into vortices which carry the pollen grains onto the pollination droplets on top of the ovules. This implies that anemophilous pollination may not be left entirely to chance (Niklas 1985).

POLLEN VECTORS			SPECIFICITY	EFFICIENCY	MAXIMUM DISTANCES REACHED BY STILL VIABLE POLLEN
animal (zoophilous)	insects (entomophilous)	bees	high	very high	several hundred m
		wasps	moderate	moderate	a few hundred m
		flies	low	moderate	a few hunfred m
	birds		moderate	moderate	many hundred m
	small mammals		low	low	several hundred m
winds (anemophilous)	breezes		none	low	a few hundred m
	stormy winds		none	very low	hundreds of km
water (hydrophilous)	fresh water		none	low	several m
	sea water		none	low	a few hundred m

Table 1. Table depicting the vectors of pollination, their specificity, efficiency and the relative distances reached by pollen grains. Bees are the best vectors because of their specificity and efficiency. Abiotic vectors have no specificity and low to very low efficiency.

	ENTOMOPHILOUS	ANEMOPHILOUS
HABITUS	herbs > trees	trees > herbs, both living in groups, social plants
ENVIRONMENT	tropical > temperate (especially during springtime and summer	cold > temperate
INFLORESCENCES	all kinds	often aments, flowers blooming contemporaneously
FLOWER	with nectaries and scent, often showy	inconspicuous, without nectaries, often monoecious
	anthers inside the corolla	anthers outside the corolla
	stigma with small receptive surface	stigma with a large receptive surface
POLLEN	ornamented exine	smooth exine
	abundant pollenkitt	pollenkitt usually absent
	pollen grains adhere together by several mechanisms: pollenkitt, viscin, poliads, massulae, pollinia	pollen disperse separately
	pollen viability from hours to a few days	pollen viability from minutes to hours
	big pollen grains (60-300 /um)	small pollen grains (less than 60 /um)

Table 2. Peculiarities of species with anemophilous and entomophilous pollination.

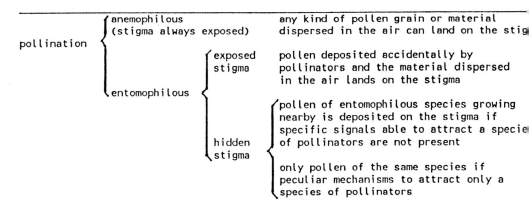

Table 3. Main pollination types, flower morphology and pollen types landing on stigma.

	number	wind	rain	birds	wind & insects	flies	midges	bees	wasps
SELF-COMPATIBLE	24	12.50		4.16		12.50		29.16	4
SELF-INCOMPATIBLE	43	30.02		2.32		2.32	2.32	55.55	
DIOECIOUS	24	54.16	4.16		4.16			4.16	3

Table 4. Percentages of pollination syndromes in 91 self-compatible self-incompatible cultivated species (Data from Frankel & Galun 1977).

	monads	monads grouped by pollenkitt or thyphine	monads grouped by viscin threads	filamentous pollen	tetrads	tetrads grouped by viscin threads or pollenkitt	poliads	massulae	pollinia
ANEMOPHILY	●				●				
HYGROPHILY	●			●					
ORNITHOPHILY		●	●			●	●		
ENTOMOPHILY		●	●		●	●	●	●	●

Table 5. Types of pollen dispersal unit and pollination syndrome.

Any kind of pollen grain in the air may fall on the stigma of anemophilous species. A limited number of grains falls on the stigma of entomophilous species (Tab 3). This number is lower if the stigma is hidden inside the flower or if the plant has peculiar devices able to attract a specific pollinator. This number is similar to the number of ovules in species with compound pollen, such as *Acacia* poliads (Knox & Kenrick 1983), or massulae or pollinia like in Asclepiadaceae and Orchidaceae.

Pollination in angiosperms is preferentially anemophilous or entomophilous. Anemophily decreases from the poles to the tropics; entomophily decreases or is restricted to warm seasons from the tropics to the poles. Hydrophily exists in a limited number of species, most of which are seaweeds (Pettitt et al 1981). Zoophily is mainly due to insects (entomophily), but there have been several reports of pollination by bats, small rodents, small birds, lizards and even monkeys.

Grasses are almost totally anemophilous, whereas orchids are almost totally entomophilous, some even species specific. Between these two extremes there are many species sharing both pollination types. *Erica arborea* is entomophilous at the onset of anthesis; bees are attracted by nectar, present in a disc at the base of the ovary, and pollen from the anthers hidden inside the corolla. This phase lasts one or two days, after which the anther filament lengthens and any pollen grains still present are dispersed on the wind (unpublished data).

Good pollinators have many hairs or bristles to which pollen can adhere and they do not damage the flower when they visit it. The mouth of the insect is very important: chewing insects such as coleoptera damage the floral parts; sucking insects such as diptera, lepidoptera and hymenoptera are the best pollinators. Good pollinators do not damage the pollen during transport and pollinate quickly. Wingless insects like ants are not good pollinators and secrete substances which kill pollen (Beattie et al. 1985). Some moths visit 25 flowers per minute (Faegri & Van der Pijl 1979). Good pollinators should also be attracted by specific signals sent by the flower (Faegri & Van der Pijl 1979). Among insects, bees are the best pollinators because they visit a single species for as long as it blooms. Being social insects, they collect more material than they need. They are predominantly sensitive to visual messages and can carry considerable loads of viable pollen. They work over a range of weather conditions and in different climates.

Two kinds of bee, polliniferous and nectariferous, visit flowers (Gould & Gould 1988). The former actively collect pollen, some of which is dropped on the stigma in passing; the latter become dusted with pollen while collecting nectar and incidentally brush the stigma. This means that a flower or inflorescence may be visited at different stages of flower receptivity by insects attracted by different rewards offered by the flower: loading and unloading of pollen is facilitated in this way. The mean life of a honey bee is 4-5 weeks and it is not until the age of three weeks that it starts to collect food; it has a range of up to 10 Km from the hive (Gould & Gould 1988). Pollen collectors seem the best pollinators for some species (Jay 1986), nectar foragers for others (Estes et al. 1983). The honey bee, *Apis mellifera*, is of considerable economic importance. It pollinates a large variety of cultivated species in different environments (Frankel & Galun 1977). The species *Apis mellifera* originated in Europe and north Africa and now has world wide distribution (Crane & Walker 1984). Bees accumulate twenty-five times more nectar than they need and each one can visit several hundred flowers per day (Faegri & Van der Pijl 1979).

A species will be visited by polliniferous and/or nectariferous bees depending on: a) how rewards are advertised; b) the type of reward: there is always pollen but not always a nectary; c) the needs of the hive; d) the kind of plants with rewards, and the type of reward, growing near the colony.

Wasps, like bees, are sensitive to specific messages but are not social insects and because they do not accumulate a lot of material, they are of less economic importance. Flies have a very low specificity and are usually sensitive to olfactory stimuli. Butterflies are extremely unspecific.

Attracting mechanisms and rewards

Faegri & Van der Pijl (1979) distinguish first and second order attractants. First order attractants include pollen, nectar, honeydew, oils produced by the elaiophore, protection against adverse conditions or for mating, a place to lay eggs. Another first order reward described by Armbruster (1984) is resin produced by floral parts used by insects in nest construction. Second order attractants include smell or floral aroma (Robacker 1988), visual messages, warmth. These mechanisms generally do not act separately but in combination according to species. They may provide a real reward or mislead pollinators (Dafni 1987). Some species of orchid attract pollinators of only one sex by chemical or morphological mimesis of the opposite sex. A male insect is attracted by the labellum of the flower which simulates the female abdomen; the flower may emanate a volatile compound very similar to the male/female pheromone (Borg-Karlson 1990). Some species of the family Araceae and the genus *Stapelia* imitate carrion and dung not only in smell (Robacker et al. 1988) but sometimes even in shape. In all cases the result of attraction by a misleading message is pollination. The insect is repeatedly deceived and remains unaware of the fact.

Bees are mainly attracted to flowers by the pollen and nectar; other insects mostly respond to nectar and other rewards. Nectar production is costly in terms of energy: 4-37% of the daily photosynthate during flowering is allocated to nectar sugar production in *Asclepias syriaca* (Southwick 1984). The removal of nectar increases nectar production but affected seed production in *Blandfordia nobilis* (Pyke 1991). The shape and position of the nectaries as well as the chemical composition of nectar have an adaptive significance (Baker & Baker 1983). Most flowers provide nectar for only part of the day, sometimes compatible with insect availability. Owing to the high energy cost of nectar, nectaries are often positioned so as to exclude non pollinating nectar collectors. Long tubular corollas is one such device. Part of the pollen produced by the flower is also consumed by pollinators. On the other hand, some species of different families have evolved a specialized polliniferous reward. The flower has two kinds of anthers: one with real pollen able to give rise to pollen tubes and to fecundate, the other to feed pollinators, usually bees. In *Lagestroemia indica*, feeding pollen is bicellular but emits only a short pollen tube (Pacini & Bellani 1986).

Pollination specificity

Flowers have a particular shape according to their pollinator. Honeysuckle (*Lonicera* sp.) has a very long corolla with nectaries at the base; it is pollinated by butterflies which transfer pollen with the proboscis while collecting nectar deep down in the corolla. Flowers seem to have universal messages to advertise nectar, because bumblebees with their short sucker succeed in robbing nectar from outside the honeysuckle, avoiding pollination. Entomophilous species growing far from their natural environment succeed in attracting insects belonging to a different fauna. Some species of *Banksia* attract different classes of pollinators such as birds, mammals and insects (Paton & Turner 1985, Ramsey 1988), but their effectiveness is different. Honey bees collecting pollen are more likely to cause pollination than birds or small mammals collecting nectar and only contacting stigmata when arriving or leaving an inflorescence (Ramsey 1988).

Several cases of specificity also exist. They mostly rely on deceiving the insect with misleading messages (Dafni 1987) and may be more specific than a real reward. Studies of reproductive ecology have shown that some species share the same pollination syndrome as a more abundant sympatric species, imitating its attraction mechanisms (Lamont 1985).

Pollen loading and unloading

Pollen grains adhere to the body of an insect when it visits a flower. During flight, the insect becomes electrostatically charged so that it collects pollen without even touching the anthers (Corbet et al. 1982, Erickson & Buchman 1983, Chaloner 1986). Pollen may leave the insect body in the same way when attracted by the stigma (Corbet et al. 1982, Erickson & Buchman 1983, Chaloner 1986). Sculptured pollen seems better adapted to becoming charged than unsculptured pollen (Chaloner 1986). Polliniferus bees transport thousands of pollen grains to the hive on their third pair of legs. Pollen grains adhering to other parts of the body of polliniferous and nectariferous bees are responsible for pollination. Explosive opening of flowers is a way in which pollinators may be dusted with pollen without damaging the flower. This syndrome occurs in several families; the animal dusted may be birds and certain insect groups (Davis 1987). Different pollen dispersing units are known, some of which match the pollination syndrome (Tab 5). Pollen grains of anemophilous species are dispersed individually. The pollinium or dispersing unit of some orchids and Asclepiadaceae contains up to several thousand pollen grains which can be transported in one flight. The case of the Asclepiadaceae is unique because the pollinium must be positioned by the pollinator in the right way otherwise germination does not occur (Kunze 1991). In all other cases, pollen grains fall randomly on the stigma and are able to germinate.

Pollination and incompatibility

Different kinds of pollen can land on the stigma of anemophilous and entomophilous species (Tab 3). Comparing the pollination syndromes and compatibility in 91 cultivated species (Tab 4) we find that there are 24 self compatible species, all of which are hermaphrodite, and a high percentage are pollinated by several species of insect. A relatively low percentage is pollinated by bees. This aspecificity may occur because self pollination is possible. There are 43 self-incompatible species, 27 of which are monoecious and the others hermaphrodite. More than 50% are pollinated by good pollinators such as bees; 13 species are pollinated by wind, 11 of these are grasses and two gymnosperms. Self incompatibility does not exist in dioecious species because male and female individuals have different genotypes. Few of them are entomophilous because male and female flowers must attract pollinators with similar messages and this occurs in only a few instances.

Although wind pollination is "expensive", it suits this type of plant best; in fact, more than 50% are wind pollinated.

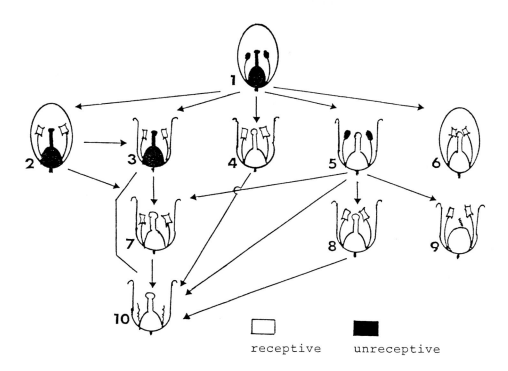

receptive   unreceptive

Fig. 1. Representative sequences of anthesis in a bisexual flower. Flower anthesis does not occur in the cleistogamous flower (6), pollen grains germinate inside the anther and pollen tubes reach the ovule through stigma and style. The anther may open when the flower is closed (2); they usually open after flower anthesis (3,4,8 and 9). Male and female receptivities can overlap (pathways 2-3-7-10, 2-7-10, 3-7-10, 4-10, 5-7-10, 5-8-10) or otherwise (pathways 3-10, 5-9). In the last two cases the receptivity of one sex is triggered by the loss of receptivity of the other. In some proterogynous species (5), if no foreign pollen lands on the stigma there are two possible strategies for self pollination: a) filaments move and the anther opens close to the receptive stigma causing self pollination (8); b) stigmatic lobes move towards anthers when they dehisce.

References

Armbruster WS (1984) The role of resin in angiosperm pollination: ecological and chemical considerations. Amer J Bot 71: 1149-1160

Baker HG, Baker I (1983) Floral nectar sugar constituents in relation to pollination type. In "Handbook of experimental pollination biology" (Jones CE, Little RJ eds) pp. 117-141. van Nostrand Reinhold Co Inc, New York

Beattie AJ, Turnbull C, Hough T, Jobson S, Knox RB (1985) The vulnerability of pollen and fungal spores to ant secretions: evidence and some evolutionary implications. Amer J Bot 72: 606-614

Bonner LJ, Dickinson HG (1989) Anther dehiscence in *Lycopersicum esculentum* Mill. I. Structural aspects. New Phytol 113:97-115

Borg-Karlson A-K (1990) Chemical and ethological studies of pollination in the genus *Ophrys* (Orchidaceae). Phytochemistry 29:1359-1387

Chaloner WG (1986) Electrostatic forces in insect pollination and their significance in exine ornament. In "Pollen and spore: form and function" Blackmore S, Ferguson IK (eds) pp. 103-108, Academic Press, London

Clifford SC, Owens SK (1988) Post-pollination phenomena and embryo development in the Oncidiinae (Orchidaceae) In "Sexual reproduction in higher plants" Cresti M, Gori P, Pacini E (eds) pp. 407-412, Springer Verlag, Berlin

Corbet SA, Beament J, Eisikowitch D (1982) Are electrostatic forces involved in pollen transfer? Pl Cell Environ 5:125-129

Crane E, Walker P (1984) "Pollination Directory for World Crops". International Bee Research Association, London

Dowding P (1987) Wind pollination mechanisms and aerobiology. Inter Rev Cytol 107:421-437

Erickson EH, Buchmann SL (1983) Electrostatics and pollination. In: Handbook of experimental pollination biology. Jones CE, Little RJ (eds) pp 173-184

Dafni A (1987) Pollination in orchids and related genera: evolution from reward to deception. In: "Orchid Biology" Arditti J (ed) pp 80-104, Comstock Publishing Associates, Ithaca

Davis MA (1987) The role of flower visitors in the explosive pollination of *Thalia geniculata* (Maranthaceae), a Costa Rican marsh plant. Bull Torrey Bot Club 114:134-138

Erickson EH, Buchmann SL (1983) Electrostatic and pollination. In "Handbook of Experimental Pollination Biology" Jones CE, Little RJ (eds) pp. 173-184. van Norstrand Reinhold, New York

Estes JR, Amos BB, Sullivan JR (1983) Pollination from two perspecitves: the agricultural and biological sciences. In "Handbook of Experimental pollination biology" Jones CE, Little RJ (eds) pp. 173-183, van Norstrand Reinhold Co Inc, New York

Faegri K, Pijl Van der L (1979) The principles of pollination ecology. Pergamon Press, Oxford

Frankel R, Galun E (1977) Pollination mechanisms, reproduction and plant breeding. Springer-Verlag, Berlin

Gould JL, Gould CG (1988) The honey bee. Scientific American library, New York

Jay SC (1986) Spatial management of honeybees on crops. Ann Rev Entomol 31:49-65

Keijzer CJ (1987) The process of anther dehiscence and pollen dispersal. II. The formation and the transfer mechanism of pollenkitt, cell wall development of the loculus tissues and function of the orbicules in pollen dispersal. New Phytol 105:499-597

Knox RB, Kenrick J (1983) Polyad function in relation to the breeding system of *Acacia*. In "Pollen: Biology and implication in plant breeding" Mulcahy DL, Ottaviano E (eds) pp.411-417, Elsevier Science Publ Co, New York

Kunze H (1991) Structure and function in asclepiad pollination. Pl Syst Evol 176:227-253

Lamont B (1985) The significance of flower color change in eight cooccurring shrub species. Bot J Linn Soc 90:145-155

Niklas KL (1985) The aerodynamics of wind pollination. Bot Rev 51:328-386

Pacini E (1990a) Tapetum and microspore function. In "Microspores: evolution and ontogeny" Blackmore S, Knox RB (eds), Academic Press, London

Pacini E (1990b) Harmomegathic characters of Pteridophyta spores and Spermatophyta pollen. Pl Syst Evol (supp. 5) 53-69

Pacini E, Bellani LM (1986) *Lagestroemia indica* L.: form and function. In "pollen and spores: form and function" Blackmore S, Ferguson IK (eds) pp. 347-357, Academic Press, London

Paton DC, Turner V (1985) Pollination of *Banksia ericifolia* Smith: Birds, mammals and insects as pollen vectors. Aust J Bot 33:271-286

Perryman TC, Marcellos H (1988) The rhythm of flower opening in *Vicia faba* L.. FABIS Newsletter 21:17-18

Pettitt JM, Ducker SC, Knox RB (1981) Submarine pollination in seagrasses. Sci Am 244:135-143

Pyke GH (1991) What does it cost a plant to produce floral nectar? Nature 350:58-59

Pleasants JM, Chaplin SJ (1983) Nectar production rates of *Asclepias quadrifolia*: causes and consequences of individual variation. Oecologia 59: 232-238

Robacker DC, Meeuse BJD, Erickson EH (1988) Floral aroma: how far will plants go to attract pollinators. Bioscience 38: 390-398

Ramsey MW (1988) Differences in pollinator effectiveness of birds and insects visiting *Banksia menziesii* (Proteaceae). Oecologia 76: 119-124

Southwick EE (1984) Photosynthate allocation to floral nectar: a neglected energy investment. Ecology 65:1775-1779

Stelleman P (1984) Reflections on the transition from wind pollination to ambophyly. Acta Bot Neerl 33:497-508

Pollen Capture, Adhesion and Hydration

J.S. Heslop-Harrison
Karyobiology Group
Department of Cell Biology
JI Centre for Plant Science
Norwich NR4 7UJ England

Introduction

Pollen capture by the stigma initiates a complex sequence of events that can lead to the hydration and germination of the pollen grain, and eventually to penetration of the pollen tube and fertilization. This chapter will review aspects of the physiology and physics of events related to pollen capture, hydration and adhesion, but not concepts related to the physical capture of pollen by the stigma. Work on the aerobiology of pollen capture in species pollinated by wind is reviewed by Niklas (1987), and in water pollinated species by Ducker and Knox (1976). Many ecological works discuss aspects pollen transfer animal vectors.

Pollen dehydration occurs during the final stages of maturation within the anther, and presumably continues up to and after dehiscence. Since the mature pollen is almost inactive metabolically (Hoekstra and Bruinsma 1980), hydration levels may fluctuate with changes in atmospheric conditions. Once the pollen lands on a stigma, rehydration can occur very rapidly, leading to regaining of active metabolism: in rye (*Secale cereale*), hydration starts within a few seconds of a pollen grain landing on the stigma, and in *Brassica*, germination often occurs within ten minutes of pollen capture. Study of fast events, particularly when they involve an interaction between the pollen and stigma, is difficult. Some aspects of hydration and the early events preceeding germination can be studied *in vitro*, by placing pollen in liquid media or on semi-solid media, but active interactions must be studied on the stigma.

Dehydrated and Hydrated Pollen Structure

Dehisced pollen grains have widely varying hydration status between species, ranging from 6% water content in *Populus* to 60% in *Zea mays*. Because all routinely used biological fixatives are aqueous, and hence change hydration status, it is difficult to study dehydrated pollen by cytochemical methods or electron microscopy. However, an understanding of the events of dehydration - and rehydration - is important for fundamental research and for practical applications, since the understanding will help to overcome barriers to hybridization or limitations of fertility caused by problems associated with rehydration, and find new methods for the long term preservation of germplasm or breeding materials.

The thickened and often sculptured pollen wall, the exine, their high lipid content and the refractile nature of most pollen grains, prevent useful imaging inside the grain using light microscopy. Conventional fixation of hydrated and unhydrated grains in glutaraldehyde, for electron microscopy, indicated that there are substantial differences in internal structure between grains plunged into fixative and those hydrated before fixation (Heslop-Harrison 1979). More recently, vapour phase and cryo-fixation methods have been used for electron microscopy of dehydrated and hydrating pollen without the possibility of rehydration, and physical imaging methods have examined membranes within grains.

Anhydrous vapour fixation of unhydrated, partially and fully hydrated grains of *Brassica* was used by Elleman and Dickinson (1986) to show the change in pollen ultrastructure during hydration. Grains were placed in a closed jar containing osmium tetroxide vapour, that pre-fixed the grains without changing hydration, before aqueous fixation for electron microscopy. A previously undetected, membrane-like coating superficial layer (CSL) was found around the pollen grains, which is important since this structure is the first to contact the stigmatic pellicle. However, striking changes in cytoplasmic organization were particularly noteworthy providing remarkable evidence that the membranes changed from a micellar form to a hydrated, continuous form in a few minutes after hydration.

Tirawi, Polito and Webster (1990) used freeze substitution to look at unhydrated (7% moisture content) and hydrated viable pollen of pear in the electron microscope. Grains were quench frozen in liquid propane in an unhydrated state or after 2 min in a germination medium, before freeze-substitution with 1% osmium tetroxide in acetone, and embedding in resin. The data showed that the ultrastructure of the hydrated pollen grains was extremely different from that of dry pollen. Dry pollen showed a shrunken cell morphology: there were infoldings of the wall and a crowding of the organelles. However, there were vesicles associated with the extracytoplasmic, as well as the cytoplasmic, surface of the plasma membrane, and many multilamellate membranes and dense osmiophilic bodies in the cytoplasm, that were presumably associated with the dehydrated status but not simple shrinking. In the grains that were hydrated for two minutes, the cell wall infoldings, intine wall corrugations, osmiophilic bodies and the multilamellate membranes disappeared, and the grains showed an ultrastructure generally typical of plant cells. The authors suggest that the complex, tightly packed, multilamellate membrane conserves membrane material, which become surplus as the grain dehydrates, in a form that is ready rapidly to reform the membrane system of the cell upon rehydration.

Crowe, Hoekstra and Crowe (1989) were able to examine the state of membranes within bulk samples of pollen by Fourier transform infrared spectroscopy. Their results demonstrated a significant change in the structure of the membrane between the hydrated and dehydrated state. Using pollen from *Typha latifolia*, they showed that the membrane phospholipids in the dry pollen were in a gel phase (dry bilayer). If a grain is then placed into water (or a medium), the gel phase phospholipids change to a liquid crystalline phase (hydrated bilayer), and the grain contents leak through the membranes during the phase transition. The cytoplasmic leakage causes pollen death. However, when a grain is rehydrated over water vapour, or heated slightly, the gel phase lipids undergo the transition to the liquid crystalline phase without leakage. When the grain is subsequently placed into water, there is no leakage of the grain contents through the hydrated bilayer in the liquid crystalline state, so the grain remains viable and can germinate. Thus it is vital for the dry pollen to hydrate under conditions that allow the membrane phase change to occur without leakage of the grain contents for viability to be maintained.

The model of pollen hydration proposed by Heslop-Harrison (1979) suggested that the plasma membrane is porous and ineffective at the time of dispersal so it does not provide any osmotic barrier. Membrane integrity is restored by rehydration on the stigma after pollination. The data discussed above strongly support this model. The hydration of the pollen grain either on a stigma or on an artificial medium is effected by the passage of water through the germination apertures of the grain, where the intine is more accessible. The apertures themselves often show refined adaptations for the regulation of water loss and uptake (Heslop-Harrison, Heslop-Harrison and Heslop-Harrison 1986). If rehydration is too rapid, considerable imbibitional leakage of the cytoplasm occurs, leading to death of the pollen. In the dehydrated grains of many species, including hazel and rye, the apertures are sealed either by 'lids' of by infolding of the lip-like markings of germinations slits, which is completed by the coalescences of the surface lipids (Heslop-Harrison et al 1986; Heslop-Harrison 1987). Contact with a moist surface leads to dispersal of the lipid, exposing the underlying intine that then allows hydration. As the intine hydrates, the slits gape more and water intake is further facilitated.

Stigma Interactions

The work above discusses the changes in pollen upon hydration *in vitro*, and similar changes occur when the grain hydrates on a stigma. As discussed by Heslop-Harrison (1987), successful pollination requires a high level of co-adaptation between the pollen and the stigma and style. At the time of pollen capture and during germination, the interactions involve the female sporophyte and the male gametophyte, with surface, paternally derived, materials carried by the pollen grain. Breakdown of precise co-adjustment at any point will act as a fertility barrier. Because of the randomness associated with most pollination systems, stigmata commonly capture pollen from foreign species, and this is generally rejected. There can be crude maladjustments - size of pollen and stigma papillae, or water tensions in the stigma and pollen can prevent initial grain hydration, or cause osmotic bursting of the pollen - although specific interactions are also important (Heslop-Harrison and Heslop-Harrison 1975).

Attachment to the Stigma

In species with the dry type of stigma, the capture of pollen depends upon its attachment to the surface of the papillae, but the attachment is apparently more complex than simple adhesion. Maize pollen transferred to the stigma is held, but easily dislodged. During the first 3-5 minutes, the adhesion increases in strength until germination of the grain. Ferrari et al (1985) demonstrated the several stages of binding in compatible combinations in *Brassica*. The first "force", begining seconds after pollen contact with the stigma, was dissociated in methanol or sodium hydroxide, but not water. A second binding force, which was dissociated only in sodium hydroxide, occurred later, after germination, a third, sodium hydroxide stable, force bound the germinating grain to the stigma. They discuss their results in terms of van der Waals forces and lipid polymerization, although such forces would tend not to give the specificity shown by the work of Salker, Elleman and Dickinson (1988).

Hydration on the Stigma and Recognition Reactions

In many angiosperms, pollen from the same or genetically identical individuals is rejected - the plant is self-incompatibile. In all sporophytic and some gametophytic systems (see chapters by de Nettancourt and Thompson), the stigma surface is the site of self-incompatibility responses, and it is now clear that the compatibility recognition events can occur even before the pollen is fully hydrated, at least in *Brassica*. Indeed, the initiation of hydration of the grain by passage of water from the stigma, and the transfer of pollen wall proteins from the pollen exine and intine to the stigma surface occur simultaneously (Heslop-Harrison and Heslop-Harrison 1985). However, the pollen-stigma interactions are complex, and environmentally modified, so present descriptions are incomplete.

In compatible *Brassica* pollinations, hydration of the grain on the stigma is often complete within five minutes following capture of the grain, and in this period, emissions from the pollen wall sites are transferred to the stigma surface (Heslop-Harrison 1975). In compatible combinations, germination begins in 3-10 minutes and the tube penetrates the cuticle in 10-30 min. In an incompatible reaction, the rejection responses are first evident in 5-10 min.

Zuberi and Dickinson (1985) assayed the hydration of *Brassica* pollen grains by measuring their length to width ratio, which is related to the grain volume and hence hydration state. The pollen grains could take up water from a humid atmosphere as well as the stigma, but long periods (3 h) on a slide at high humidity lead to the pollen becoming inviable. On the stigma, dehydrated pollen immediately started hydrating, and, in compatible pollinations, the process continued until germination had taken place. However, the rate of hydration depended on the S genotypes of both the pollen and stigma, as well as the compatibility between the two. In humid (low vapour pressure deficit) conditions, pollen tube organization and penetration were adversely affected. Although the pollen tube usually germinated, its growth tended to become disorganized after emergence from the grain and it was not able to penetrate the papillar surface. Thus, the grain can obtain water both from the stigma and from the atmosphere, and some early parts of the incompatibility reactions (leading to rejection of the pollen by the stigma) were overcome by hydration from the atmosphere. Thus they could conclude that hydration plays a key role in the operation of the sporophytic self-incompatibility system in *Brassica*.

Control of Hydration

The regulated passage of water from stigma to pollen is an essential prerequisite for successful pollen tube development. Heslop-Harrison (1979) advanced a hypothesis that slow hydration by water from the atmosphere and the stigma enables the developing pollen grain to initiate the process of cytoplasmic reorganization that is crucial for normal germination and penetration of the stigma, and then gave a theoretical consideration of the water potentials of the different stages of hydration. Elleman and Dickinson (1986) presented data to confirm the model: the dry pollen first extracts water from the stigma by matric potential, and then, after formation of an intact plasm membrane, by a system based on turgor pressure differentials. Exudation of water from the grain can occur at an intermediate stage. The water flow is regulated by a balance between the hydraulic pull exerted by the pollen grain and the resistance to flow provided by the stigma. The structures involved in this process, reviewed by Sarker, Elleman and Dickinson (1988), are complex: the pollen is bounded by a lipidic coating invested by a membrane-like coating superficial layer (CSL; see above), while the stigma surface consists of an enzyme rich superficial pellicle investing an irregular cuticular layer. The papilla boundary is well adapted for the regulated passage of water to the pollen through its cellulosic wall with microchannels, and osmotic changes may open cracks in the cuticle to enable more rapid water flow, and later penetration of the pollen tube.

The events within the pollen grain following hydration are the subjects of other chapters in this volume (see Pierson, Cresti). These events are activated and occur rapidly after hydration, but clearly stigma attachment and the precise regulation of hydration is an important event in initiation of the pollen stigma interaction in many species.

References

Crowe JH, Hoekstra FA, Crowe LM (1989) Membrane phase transitions are responsible for imbibitional damage in dry pollen. Proc Natl Acad Sci USA 86:520-523.

Ducker SC, Knox RB (1976) Submarine pollination in seagrasses. Nature 263:705-706.

Ferrari TE, Best V, More TA, Comstock P, Muhammad A, Wallace DH (1985). Intercellular adhesions in the pollen-stigma system: Pollen capture, grain binding, and tube attachments. Am J Bot 72:1466-1474.

Gaget M, Saied C, Dumas C, Knox RB (1984). Pollen-pistil interactions in interspecific crosses of *Populus* (sections Aigeiros and Leuce): Pollen adhesion, hydration and callose responses. J Cell Sci 72:173-184.

Heslop-Harrison J (1975) Male gametophyte selection and the pollen-stigma interaction Gamete competition in Plant and Animals. DL Mulcahy. North Holland, Amsterdam.

Heslop-Harrison J (1979) An interpretation of the hydrodynamics of pollen. Am J Bot 66:737-743.

Heslop-Harrison J (1987) Pollen germination and pollen-tube growth. Int Rev Cytol 107:1-78.

Heslop-Harrison J, Heslop-Harrison Y (1985) Surfaces and secretions in the pollen-stigma interaction: A brief review. J Cell Sci Suppl 2:287-300.

Heslop-Harrison J, Heslop-Harrison Y 1975. Ann Bot 39:163-165

Heslop-Harrison Y, Heslop-Harrison JS, Heslop-Harrison J (1986) Germination of *Corylus avellana* L. (Hazel) pollen: Hydration and function of the oncus. Acta Bot Neerl 35:265-284.

Hoekstra FA, Bruinsma J (1980) Control of respiration of binucleate and trinucleate pollen under humid conditions. Physiol Plant 48:71-77.

Knox RB, Williams EG, Dumas C (1986). Pollen, pistil and reproductive function in crop plants. Plant Breeding Rev 4:9-79.

Niklas KJ (1987) Pollen capture and wind-induced movement of compact and diffuse grass panicles: implications for pollination efficiency. Am J Bot 74:74-89.

Sarker RH, Elleman CJ, Dickinson HG (1988) Control of pollen hydration in *Brassica* requires continued protein synthesis, and glycolsylation is necessary for intraspecific incompatibility. Proc Natl Acad Sci USA 85:4340-4344.

Stead AD, Roberts IN, Dickinson HG (1979) Pollen-pistil interactions in *Brassica oleracea*: Events prior to pollen germination. Planta 146:211-216.

Tiwari SC, Polito VS (1988) Spatial and temporal organization of actin during hydration, activation and germination of the pollen in *Pyrus communis* L.: a population study. Protoplasma 147:5-15.

Tiwari SC, Polito VS, Webster BD (1990) In dry pear (*Pyrus communis* L.) pollen, membranes assume a tightly packed multilamellate aspect that disappears rapidly upon hydration. Protoplasma 153:157-168.

Zuberi MI, Dickinson HG (1985). Pollen-stigma interaction in *Brassica* III. Hydration of the pollen grains. J Cell Sci 76:321-336.

POLLEN TUBE EMISSION, ORGANIZATION AND TIP GROWTH

M. Cresti, A. Tiezzi
Dipartimento di Biologia Ambientale
Università di Siena
Via P.A. Mattioli, 4
53100 Siena (Italy)

When the flower open and anther dehisces, the mature pollen is liberated and transported by different agents (principally wind and insect) to the stigma. According to the phisyological condition of the stigma (dry, wet) the germination of pollen grain take place.

What makes pollen on expecially attractive experimental material is the fact that many pollen grains are able to germinate and grow (figs 1, 2, 3) in the laboratory on relatively simple artificial growth media. In addition, many pollen species retain their viability for several weeks or months when refrigerated or frozen.

Pollen growth media usually contain sucrose as the carboydrate source, while boric acid and calcium are supplimented to stimulate germination and growth. Various other trace elements such as zinc, manganese and iron are required for germination of certain pollen species (see Heslop- Harrison, 1987).

The mature and ungerminated pollen grain is a dehydrated and metabolically inactive structure. Using rapid freezing and substitution technique the ultrastructural characheristics of unhydrated pollen has been observed (fig. 4). The first phase in pollen germination is the hydration and imbibition of water followed by the activation. The germination results in the outgrowth of the pollen tube. Between the hydration and germination, many morphological and biochemical changes, e.i. protein synthesis, take place. The main morphological changes concern the rough endoplasmic reticulum (RER) that in the ungerminated and unactivated grain, such as *Lycopersicum* and *Nicotiana*, is aggregated in stacks (fig. 5). During the activation, the RER cisterns setting free from the stacks and become free in the cytoplasm (fig. 6). Furthermore the dictyosomes, which are inactive in the unhydrated grain, begin to actively produce vesicles, expecially in proximity to the germination pore.

The growth of the pollen tube is different from that of most other plant cells where the growth takes place over the entire surface of the cell. However the pollen tube exention occurs only in the tip region and this was first evidenced in *Lilium* (Rosen 1964). Using carbon powder attached to the pollen tube it was demostrated that the growing zone is restricted to the tip (3 - 5 /um). In this respect the pollen tube growth is similar to that of root hair and fungal hyphae (Steer and Steer, 1989).

Electron microscopy studies have show that in many species a characteristic functional zonation exists along the length of the pollen tube (Cresti et al., 1977). The growing zone is restricted at the tip and generally contains only vesicles (of different types according to the species), a sub-apical zone rich in organelles but particularly Golgi producing vesicles, a wide nuclear zone containing vegetative nucleus and generative cell or sperm cell and a vacuolization and callosic plug formation zone dividing the leaving part of the tube from the inactive ones (for recent reviews see Cresti and Tiezzi, 1990; Pierson and Cresti, 1992). The active part of the pollen tube exibits an intense streaming of cytoplasm. The streaming occurs in different manner but the fountain-like appearence seem to be typical before many pollen tube species.

1. Apical zone and tip growth

During the fertilization process in all investigated angiosperm plants each pollen grain germinates only one pollen tube. All growing pollen tubes present unidirectionality of growth and the tube tip is considered the only region that undergoes growth and is also responsible for defining the diameter of the tube (Steer and Steer, 1989). The tip is bounded by a fibrillar cell wall described as pectic nature (Roggen and Stanley, 1971); the plasma membrane frequently exhibits fusion profiles with secretory vesicles membrames. The only organelles visible are the secretory vesicles present as individual vesicles (fig.7). Only in few cases single or bundles of microtubules (fig.8) or short segments has been reported (Lancelle et al., 1987). Antitubulin markers give a diffuse reaction at the tip, indicating the presence of free tubulin rather than the discrete fluorescence of microtubules (Derksen et al., 1985) whereas rhodamine-phalloidine staining put in evidence the presence of actin filaments.

Characteristically the rate of growth differs among different pollen grain species; however as the tube growth has been mainly studied by in vitro germination tests, it is unclear how much external "factors" are able to influence the modalities and rate of pollen tube growth.

It is well established that Golgi vesicles containing polysaccharide wall precursors continuously fuse at the tube tip contributing both in the elongation of the pollen tube wall and pollen tube plasmamembrane (see

Heslop-Harrison, 1987; Steer and Steer, 1989). In pollen tubes prepared with rapid freezing and substitution (Lancelle et al., 1987) the fusion and incorporation of vesicles membrane into the plasma membrane are rare, indicating that the process is very rapid. Perhaps additional investigations of the modalities of Golgi vesicles movement to the tube apex could represent a significative contribution to the study of the process of pollen tube elongation. In this view the discovery of a complex cytoskeletal apparatus inside the pollen tube (for reviews see Tiezzi, 1991; Pierson and Cresti 1992; Pierson and Li, this volume) opened new interesting perspectives. In fact the observations concerning the active role of an actin-myosin transducing system in the movement of organelles along the tube (Heslop-Harrison and Heslop-Harrison 1989) contributed in the explanation of the videomicroscopical investigations (Heslop-Harrison and Heslop-Harrison 1988; Pierson 1990) on the processes of cytoplasmic streaming and continuous cytoplasm reorganization.

In our lab we are interested to the study of the role of microtubules in the pollen tube. The pollen tube is rich in microtubules and although they have been repeatedly investigated by different techniques (Tiezzi, 1991), their function still remains substantially unclear.

By a monoclonal antibody (k71s23) obtained in our lab, we have recently identified an immunoreactive homolog of mammalian kinesin in the pollen tube of *Nicotiana tabacum* (Tiezzi et al., 1991). In different animal systems kinesin is a cytoskeletal polypeptide having ATPase activity and cooperating together with microtubules in assuring the movement of vesicles (Goldstein, 1991). By conventional and confocal immunofluorescence investigations we were able to detect the cytoplasmic domain of such polypeptide mainly in proximity of the plasmamembrane of the pollen tube tip (also the microtubules localize along the tube in proximity of the plasmamembrane). In particular the immunostaining was localized on the surface of round particles of similar diameter to the vesicles observed by the electron microscope in the same pollen tube region. If a kinesin immunoreactive homolog is present on the surface of vesicles, in analogy to that observed in animal cells, it should interact together with microtubules. But as reported by immunofluorescence (Derksen et al., 1985) and ultrastructural investigations (Lancelle et al., 1987) microtubules do not seem to be present at the pollen tube apex and consequently the possible function of the kinesin immunoreactive homolog must be carefully evaluated. For instance (i) it is possible that in plant cells such polypeptide works with different molecular modalities respect to animal systems. In this view (ii) there could be a sort of functional cooperation with actin filaments. (iii) Another possibility is that the kinesin immunoreactive homolog interacts with the depolimerized tubulin present at the tube apex. (iiii) It is also possible that the microtubules are present at the apex and that they are destroyed or depolymerized during fixation procedures. In this view recent immunofluorescent reinvestigations carryed out in our lab reinforce this hypothesys (Li et al., manuscript in preparation).

In conclusion although the mechanisms of pollen tube growth remain still substantially unclear, it is becomming evident that the movement of vesicles and consequently the pollen tube growth is mediated by the cytoskeletal machinery. The role of microtubules will be better elucidated in future; however there are concrete perspectives for an active role of microtubules in driving the vesicles to the pollen tube apex.

2. Sub-apical zone

The pollen tube wall in this zone appears double-layered, the inner layer being callosic and as thick as the outer one which is pecto-cellulosic (Ciampolini et al. 1982; Cresti et al. 1977, 1980). Generally this sector of pollen tube contains many active dictyosomes producing vesicles, abundant rough endoplasmic reticulum and occasionally some smooth endoplasmic reticulum both with dilated cisternal, possessing electron dense intracisternal material. The ribosomes are frequently aggregated as polysomes, the mitochondria are rod-shaped and the proplastids contain starch. Small vacuoles and few lipid bodies are sometime discernible. In longitudinal sections most of these organelles appear to be lying parallel to the long axis of the pollen tube but their orientation becomes more and more random as we move closer to the tip, perhaps in accordance with the direction of cytoplasmic streaming and growth.

In freeze-substituted materials, single microtubules parallel with fine filament are longitudinally arranged according to the growth direction of the tube. The role of cytoskeletal apparatus in the pollen tube has been recently reviewed (Tiezzi, 1991, Pierson and Cresti 1992, Pierson and Li 1992, this volume).

3. Nuclear zone

Although the outer part of the tube wall continues to be pecto cellulosic, the inner callosic layer become much ticker than the sub-apical zone. Both the generative cell alongwith its nucleus and the vegetative nucleus lie in this zone. The generative cell is generally spindle shaped and also highly lobed. It is surrounded by a "wall" that, after rapid freezing and substitution, appears as regular structure.

In fact the two plasma membrane (vegetative and generative) are very smooth and closely appressed to a layer of wall material. Plasmodesmata are not seen, but this does not exclude the possibility that they may exist at an earlier stage of development (Cresti et al. 1987). The generative cell has its own cytoskeletal apparatus (see Tiezzi, 1991; Palevitz and Tiezzi 1992; Pierson and Li 1992 present volume). The generative nucleus occupies the larger part of the cell, the remaining cytoplasm contains some dictyosomes, ribosome and RER cisterns, bundles of microtubules, sporadic mitochondria and very few vacuoles. Depending on the species, plastids may or may not be present. The vegetative nucleus lies very near to the generative cell; generally it is lobed. The composition of the vegetative cytoplasm of the nuclear zone present a higher number of vacuoles if compared with the sub-apical zone (for review see Cresti and Tiezzi 1990).

4. Vacuolization and callosic plug formation zone.

The thickness of pecto-cellulosic part of the pollen tube wall in this region remains unchanged but the callosic layer is much thicker than that of the nuclear zone. The cytoplasm is full of vacuoles of larger dimension. Mitochondria, ribosomes, polysomes, and dictyosomes producing vesicles occupy the tube cytoplasm. Some SER and a more large number of RER cisternae, associated with amyloplasts are also present. The early stages of differentiation of the callose plug become visible and the ingrowing callosic area is seen interspersed with the fusing callosic grains (Kroh and Knuiman, 1982; Cresti and Van Went 1976). They consist of electron-dense finely granular, fibrillar material, occasionally surrounded by a translucent halo (see Cresti and Van Went 1976).

5. Ca^{2+} gradient and pollen tube growth.

It is now fairly well established that calcium plays an important role during pollen germination and pollen tube tip growth (see Herth et al. 1991). What is more significant is the establishment of the tip-to-base gradient of both membrane-associated and free cytoplasmic calcium in the actively growing pollen tubes (Reiss and Herth, 1978; Steer and Steer, 1989). Recently, studies from this laboratory (Tirlapur and Cresti, 1992) using the technique of computer assisted video image analysis have shown that apart from the presence of a calcium gradient, there also is an overlapping gradatory distribution of the calcium receptor protein "calmodulin". It is now being

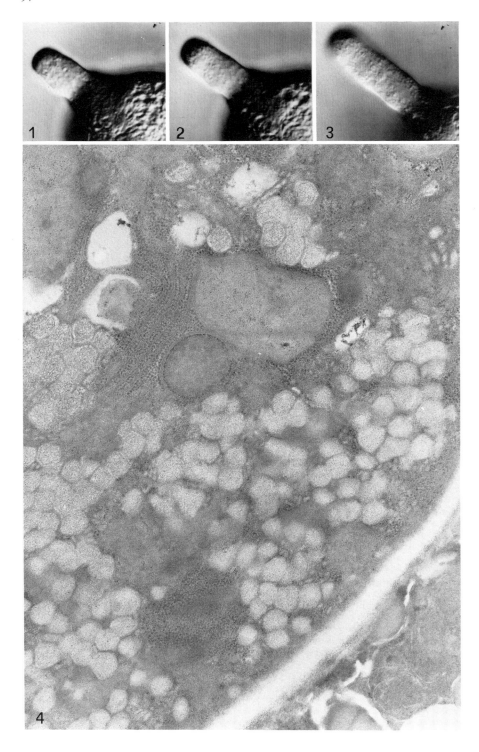

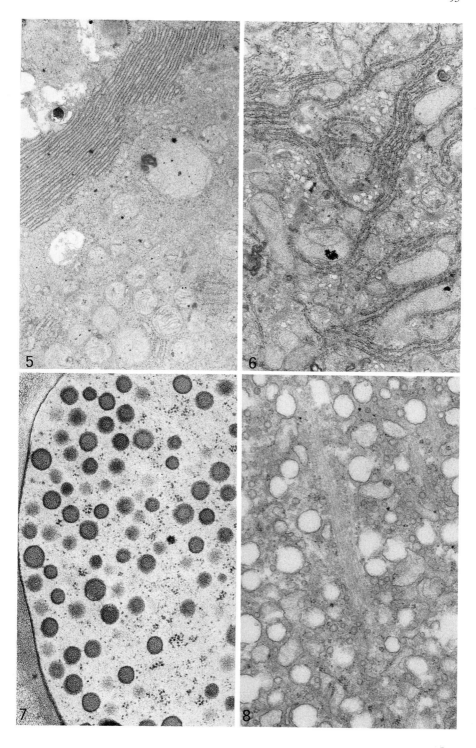

postulated that the gradient of calcium and the calcium homeostasis in general during pollen tube growth is regulated by the voltage sensitive calcium channels and by the Ca^{2+}-ATPase system present at the tube tips (see Tirlapur 1992 this volume for a discussion on this point).

Experimental studies using various calcium chelators and ionophores (Reiss and Herth, 1979,1980; Kohno and Shimmen, 1987,1988) have also shown that the cytoplasmic streaming is drastically inhibited indicating that the movement and stratification of organelles in the pollen tubes is calcium dependent. Because many of the mechanochemical motorproteins are calcium modulated it is plausible to suggest that calcium plays a pilotal role during the process or organelle movement and pollen tube growth in general.

Acknowledgements

The authors wish to thank Dr. U.K. Tirlapur for critically reading the text. This research was supported by the ECC-BRIDGE project, contract no. Biot CT90-0172 (BRMD).

Explanation of the figures

Figs. 1-3 - Video enhanced contrast differential interference contrast (VEC-DIC) microscopy of a growing pollen tube of *Ornitogalum virens*. Time lapse between 1 and 3 about 10 /um. (Courtesy of E.S. Pierson, Siena)

Fig. 4 - Portion of mature unhydrated pollen of *Arabidopsis taliana* after rapid freezing and substitution technique. The method permit to have a good presevation of the cytoplasm. x 42,500

Fig. 5 - Portion of mature hydrated pollen of *Lycopersicum peruvianum* after chemical fixation. The RER is mainly aggregated in stacks. x 21,200

Fig. 6 - Portion of mature pollen of *Lycopersicum peruvianum* during activation and before germination. RER cisterns are separate from the stacks and are dispersed in the cytoplasm. x 15,300

Fig. 7 - Portion of pollen tube tip of *Nicotiana tabacum* after freeze-fixation and substitution. Many vesicles are present in the cytoplasm. x 36,500

Fig. 8 - Portion of pollen tube tip of *Endymion non scriptus* after chemical fixation. The cytoplasm contain different tipes of vesicles and bundles of microtubules. x 29,700

References

Ciampolini F, Cresti M, Kapil RN (1982) Germination of cherry pollen grains in vitro-on ultrastructural study. Phytomorphology 32:364-373

Cresti M, Ciampolini F, Kapil RN (1984) Generative cells of some angiosperms with particular emphasis on thier microtubules. J Submicros Cytol 16: 317-326

Cresti M, Ciampolini F, Sarfatti G (1980) Ultrastructural investigations on *Lycopersicum peruvianum* pollen activation and pollen tube organization after self and cross-pollination. Planta 150:211-217

Cresti M, Lancelle SA, Hepler PK (1987) Structure of the generative cell wall complex after freeze substitution in pollen tube of *Nicotiana* and *Impatient*s. J Cell Sci 88:373-388

Cresti M, Pacini E, Ciampolini F, Sarfatti G (1977) Germination and early tube development in vitro *Lycopersicum peruvianum* pollen: ultrastructural features. Planta 136:239-247

Cresti M, Tiezzi A (1990) Germination and pollen tube formation. In: Microspore: evolution and ontogeny. S Blackmore, RB Knox (eds), Academic Press London New York pp.239-255

Cresti M, Van Went JL (1976) Callose deposition and plug formation in *Petunia* pollen tube in situ. Planta 133:35-40

Derksen J, Pierson ES, Traas JA (1985) Microtubules in vegetative and generative cells of pollen tubes. Eur J Cell Biol 38: 142-148

Goldstein LSB (1991) The kinesin superfamily. Trends in Cell Biol 1:93-98

Heslop-Harrison J (1987) Pollen germination and pollen-tube growth. Intern Rev of Cytol 107:1-78

Heslop-Herrison J, Heslop-Harrison Y (1989) Actomyosin and movement in the angiosperm pollen tube: an interpretation of some recent results. Sex Plant Reprod 2: 199-207

Hert W, Reiss HD, Hartmann E (1990) Role of calcium ions in tip growth of pollen tubes and moss protonema cells. In: Tip growth in plant and fungal cells. Heath IB (ed) Academic Press San Diego New York Boston London Sydney Tokyo Toronto pp.91-118

Kohno T, Shimmen T (1987) Ca^{2+}-induced fragmentation of actin filaments in pollen tubes. Protoplasma 141: 177-179

Kohno T, Shimmen T (1988) Accelerated sliding of pollen tube organelles along Characeae actin bundles regulated by Ca^{2+}. J Cell Biol 106: 1539-1543

Kroh M, Knuiman B (1982) Ultrastructure of cell wall and plugs of tobacco pollen tubes after chemical extraction of polysaccharides. Planta 154:241-250

Lancelle SA, Cresti M, Hepler PK (1987) Ultrastructure of the cytoskeleton in freeze-substitude pollen tubes of *Nicotiana alata*. Protoplasma 140:141-150

Palevitz B, Tiezzi A (1992) The organization, composition and function of the generative cell and sperm cytoskeleton. Intern Rev of Cytol (in press)

Pierson E, Cresti M (1992) Cytoskeleton and cytoplasmic organization of pollen and pollen tube. Intern Rev of Cytol (in press)

Pierson ES, Li Yi-qin (1992) The cytoskeleton of pollen grains and pollen tubes. In: Research in sexual plant reproduction. Cresti M, Tiezzi A (eds) Springer-Verlag Berlin Heidelberg New York London Paris Tokyo Hong Kong Barcelona (this volume)

Reiss HD, Herth W (1978) Visualization of Ca^{2+}-gradient in growing pollen tubes of *Lilium longiflorum* with Chlorotetracycline fluorescence. Protoplasma 97: 373-377

Reiss HD, Herth W (1979) Calcium ionophore A23187 affects localized wall secretion in the tip region of pollen tubes of *Lilium longiflorum*. Planta 145: 225-232

Reiss HD, Herth W (1980) The effect of broad range ionophore X537A on pollen tubes of *Lilium longiflorum*. Planta 147: 295-301

Roggen HP, Stanley RG (1971) Autoradiographic studies of pear pollen tube walls. Physiol Plant 24:80-84

Rosen WG (1961) Studies on pollen tube chemotropism. Am J Bot 481:889-895

Steer MW, Steer JL (1989) Pollen tube tip growth. New Phytol 111:323-358

Tiezzi A (1991) The pollen tube cytoskeleton. Electron Microsc Rev 4:205-219

Tiezzi A, Moscatelli A, Cai G, Bartalesi A, Cresti M (1992) Identification of an immunoreactive homolog of mammalian kinesin in *Nicotiana tabacum* pollen tube. Cell Motil Cytoskel 21: 132-137

Tirlapur UK, Cresti M (1992) Computer-assisted video image analysis of special variations in membrane-associated calcium and calmodulin during pollen hydration, germination and tip growth in *Nicotiana tabacum*. Ann Bot (in press)

Weisenseel MH, Nucciarelli R, Jaffe LF (1975) Large electrical currents traverse growing pollen tubes. J Cell Biol 66:556-567

THE CYTOSKELETON OF POLLEN GRAINS AND POLLEN TUBES

E.S. Pierson and Y.Q. Li
Dipartimento di Biologia Ambientale
Università di Siena
Via P.A. Mattioli, 4
I-53100 Siena
Italy

Abbreviations: Ca^{2+} dependent protein kinase (CDPK); N-ethylmaleimide (NEM); p-Chloromercuribenzoic acid (pCMB); rapid-freezing freeze-substitution (RF-FS).

Introduction

The cytoskeleton is assumed to be involved in many internal functions of the eukaryotic cell, e.g. organelle movement, mitotic and meiotic division, cell morphogenesis and cell growth (Dustin, 1984; Bershadsky and Vasiliev, 1988). Examination of a broad variaty of mono- and dicotyledones has revealed the presence of microtubules and/or actin filaments in the cytoplasm of male meiocytes, immature microspores, pollen grains, pollen tubes, male gametes (sperm cells) and their progenitors (generative cells) (reviews: Tiezzi and Cresti, 1990; Tiezzi, 1991; Palevitz and Tiezzi, 1992; Pierson and Cresti, 1992). Other cytoskeletal proteins, besides actin and tubulin, that have been found in pollen are myosin (Tang et al., 1989a; Heslop-Harrison and Heslop-Harrison, 1989a and b), kinesin (Moscatelli et al., 1988; Cai et al., 1992; Tiezzi et al., 1992) and profilin (Valenta et al., 1991).

Biochemical characterization of microtubules and kinesin in pollen

Microtubules are dynamic tubular structures with an apparent inner diameter of 14-19 nm and an outer diameter of 19-27 nm, which consist of dimers of alfa and beta tubulin (standard works on microtubules: Dustin, 1984; Bershadshy and Vasiliev, 1988; Warner and McIntosh, 1989; Warner et al., 1989; for plants: Kristen, 1986; for tip growing plant cells: Derksen and Emons, 1990). Various isoforms of alfa and beta tubulin, with a molecular weight of approximately 55KD, have been identified by one and two dimensional electrophoresis and immunoblotting in extracts of pollen of *Nicotiana tabacum* (Raudaskoski et al., 1987; Åström et al., 1991). Tubulins belong to a multigenic family (review: Fosket, 1989). The tubulin genes tualfa1 and tualfa3 from *Arabidopsis thaliana* have been described to be preferentially expressed in the flowers and pollen (Ludwig et al., 1988; Goddard and Wick, in progress). It is possible that the tubulin of the vegetative cytoplasm differs from that of the generative cytoplasm. In fact, in tobacco microtubules of the pollen tube are more sensitive to cold than those of the generative cell (Åström et al., 1991).

Kinesin molecules have been defined as microtubule dependent motors that move anionic beads towards the + end of microtubules and bind strongly to microtubules in vitro in the presence of AMP-PNP (in : Warner and McIntosh, 1989). In the native form, kinesin appears as a dimer of 300 to 400 KD, which is constituted of two rod shaped heavy chains (molecular weight usually 110-135 KD), and two, more globular light chains (55-80 KD) (reviews: Warner and McIntosh, 1989 and references herein; Goldstein, 1991). Using an antibody (K71S23) directed against the heavy chain of calf brain kinesin, cross reactions were revealed at 100 KD and, more weakly, at 108 KD on blots of extracts of tobacco pollen tubes (Cai et al., 1992; Tiezzi et al., 1992). The 100 KD polypeptide has a sedimentation coefficient between 8.5 and 9.5 S, similar to that of other kinesins. The polypeptide ATPase activity is stimulated in the presence of microtubules by a factor of approximately two (Cai et al., 1992).

Biochemical characterization of actin and myosin in pollen
The filamentous form of actin (F-actin) consists of identical actin monomers of globular (G) actin. The molecular weight of the single polypeptide ranges between 41 and 46 KD. The width of actin filaments, often referred to as microfilaments in the literature, is about 7-10 nm (reviews: Bershadsky and Vasiliev, 1988. for plants: Kristen, 1987; tip growing plant cells: Steer, 1990). The occurrence of actin in pollen was revealed for the first time in 1974 by heavy meromyosin decoration (Condeelis, 1974). Biochemical evidence for the presence of actin in pollen has been obtained by electrophoresis and immunoblotting, which showed bands around 45 KD (Tang et al., 1989a; Åström et al., 1991), detection of enhancement of myosin ATPase activity (Yen et al., 1992), and immunofluorescence labeling (Taylor et al., 1989; Tang et al., 1989b; Astrom et al., 1991). Gene expression has been analyzed in maturing microspores of *Zea mays* (Stinson et al., 1987).

Myosin I is a molecule with a rather globular shape and a single head, whereas myosin II, typical of striated muscles, has a long flexible tail and two pear-shaped heads. Characteristic of all myosins is their capacity to bind to actin and to show ATPase activity stimulated by actin binding (reviews on myosin: Bershadsky and Vasiliev, 1988; Hammer III, 1991). Myosin has been identified in pollen by electrophoresis and immunoblotting, showing a band at about 175 KD (Tang et al., 1989a), by measurements of ATPase activity (summarized in: Yen et al., 1992), immunofluorescence labeling (Tang et al., 1989a; Heslop-Harrison and Heslop-Harrison, 1989a and b) and inducement of movement of fluorescent actin filaments in the presence of a crude extract of pollen of *Lilium longiflorum* (Kohno et al., 1991).

Visualization of microtubules and actin filaments

Microtubules have been visualized in pollen by transmission electron microscopy of thin sections, dry cleaving preparation and immunofluorescence microscopy (reviews: Tiezzi, 1991; Pierson and Cresti, 1992). For ultrastructural research, the most detailed results on the actin skeleton have been obtained by means of RF-FS (Lancelle et al., 1986). The preparation can be combined with anti actin immunogold labeling (Lancelle and Hepler, 1989; Tang et al., 1989b). General visualization of the three-dimensional distribution of actin filaments has been achieved in pollen (e.g. Perdue and Parthasarathy, 1985; Heslop-Harrison, et al., 1986; Pierson et al., 1986; Tiwari and Polito, 1988a and 1988b; Tang et al., 1989b; Åström et al., 1991) by antiactin labeling or by staining with fluorescently labeled phalloidin or phallacidin, which specifically bind to F-actin (Tewinkel et al., 1989). The combination of microinjection of minute amounts of rhodamine-phalloidin in growing pollen tubes and sensitive detection with confocal laser scanning microscopy seems to be a promising approach to the detection of actin filaments in living pollen tubes (Zhang and Hepler, personal communication).

Organization and function of the cytoskeleton in pollen grains and pollen tubes

A simplified summary of the spatial organization of the cytoskeletal elements of pollen grains, pollen tubes, generative cell, and sperm cells is presented in figures 1, 2 and 3.

Pollen grains

Microtubules are normal components of male meiocytes and young microspores (summarized in Pierson and Cresti, 1992). However, immunofluorescence and electron microscopy studies are concordant in demonstrating that microtubules are no longer present in the vegetative cytoplasm by the time of pollen dishiscence (e.g. Cresti et al., 1977, 1986; Tiwari and Polito, 1990; Tiwari et al., 1990), while it is unclear whether the bulk of tubulin is also degraded at that time. The low water contents in pollen may be a reason for the depolymerization of microtubules. But this explanation is not very satisfying because the microtubules of the generative cell remain intact in mature and dry pollen. The phenomenon might be clarified by additional experiments, based on in situ hybridization of stage specific tubulin m-RNAs, systematical sectioning of RF-FS dry pollen by electron microscopy, and immunoblotting and immunolocalization at various stages during pollen maturation.

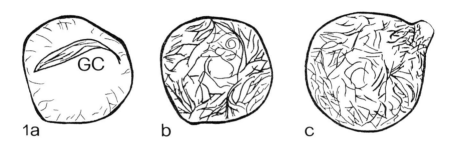

Fig. 1. Schematic representation of the organization of the microtubules during hydration, activation and germination of pollen in *Pyrus communis* L. (source: Tiwari and Polito, 1990).
a) At the beginning of pollen hydration microtubules organized in a basket-like configuration are already present in the generative cell (GC). No microtubules are found in the vegetative cytoplasm of dry pollen grains. The recovery of the microtubular skeleton initiates after rehydration of the pollen grain with the appearance of short single or V or Y-like microtuble patterns in the cortical cytoplasm. b) In a more advanced stage of activation, microtubule arrays display branching, random swirling and fan-shaped patterns. c). A concentration of small arrays of microtubules forms a collar at the base of the protruding pollen tube tip.

Tiwari and Polito (1990) applied immunofluorescence labeling to freeze-fractured pear pollen grains and clearly demonstrated that microtubules reappear in the vegetative cytoplasm as soon as the pollen is incubated in culture medium. The first patterns that they observed were short arrays and diffuse fluorescence in the cortical part of the grain (fig. 1a). In later stages of pollen activation, they found more numerous, longer and increasingly branched patterns of microtubules (fig. 1b), and finally a concentration of microtubules near the germination aperture. Eventually, microtubules formed

like a collar at the base of the emerging pollen tube (Tiwari and Polito 1988b
and 1990; fig. 1c). The polar organization of the microtubules suggests that
they are involved in the process of pollen germination. However, this process
is only slightly affected by a millimolar concentration of the anti-microtubule
drug colchicine, which is high enough to eliminate all microtubules (reviewed
in Pierson and Cresti, 1992). If microtubules do not contribute to the polar
organization of the cytoplasm and germination, which role do they play in the
pollen grain?

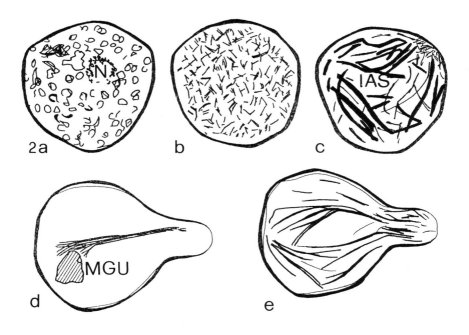

Fig. 2. Schematic representation of the organization of actin filaments during
hydration, activation and germination of pollen in *Pyrus communis* L. (source:
Tiwari and Polito, 1988a).
a). Rhodamine-phalloidin (Rh-Ph) staining in hydrated but inactivated pollen
shows numerous randomly distributed circular profiles and coarsely granular
labeling associated with the vegetative nucleus (N). b). Fusiform patterns of
Rh-Ph staining in pollen grains after 5 minutes incubation in the culture
medium. c). After 10 minutes incubation in the culture medium, coarse strands
of actin filaments are observed between the apertures (inter-apertural strands,
IAS) and short, Rh-Ph stained strands of F-actin accumulate at the apertural
pole. d). After 300 minutes of incubation in the culture medium, an actin cable
appears in close association with the male germ unit (MGU; generative cell
and vegetative nucleus). c) The other actin filaments of the vegetative
cytoplasm are oriented prepoundarilly parallel to the long axis of the emerging
pollen tube (e).

In contrast to microtubules, conspicuous masses of microfilaments or circular profiles and coarse granules of F-actin (fig. 2a) have been repeatedly shown by electron microscopy (Cresti et al., 1986) and rhodamine-phalloidin staining (Tiwari and Polito, 1988a and 1988b) in unactivated pollen of various species. During pollen hydration spicular and rod-like inclusions (Heslop-Harrison et al., 1986; Pierson, 1988; fig. 2b) or fine web-like patterns (Pierson, 1988) of F-actin have been visualized. The above described agglomerates of F-actin may be storage forms of the protein that become utilized during the course of pollen tube growth (e.g. Cresti et al., 1986), but conclusive evidence should come from concrete experiments.

The onset of germination begins with the convergence of actin filaments to the aperture (Tiwari and Polito, 1988a, 1990; fig. 2c), which is also the zone where Golgi vesicles accumulate (review: Cresti and Tiezzi, this volume). Addition of micromolar concentration of the 'anti-actin filaments' drugs cytochalasin B or D to the culture medium causes a strong inhibition of pollen germination (for example Franke et al., 1972; reviewed in Pierson and Cresti, 1992). The normal agitation and rotation movement of the organelles in the pollen grain (reviewed in Emons et al., 1991, Pierson and Cresti, 1992) is also inhibited by cytochalasins.

Pollen tubes

The subapical zone is the youngest part of the tube where microtubules are commonly found (Derksen et al., 1985). Therefore, Heslop-Harrison and Heslop-Harrison (1988) concluded that this region is most probably the site of origin of microtubules, but the precise site of assembly is not known. Coated pits and coated vesicles have often been seen before in dry cleaving preparations of pollen tubes (Derksen et al., 1985; Pierson et al., 1986). Microtubules, perhaps, play a role together with coated pits and coated vesicles in the process of internalization of plasma membrane material, as postulated by Steer and his co-workers (Steer, 1990).

Many structural investigations have documented the presence of parallel, net axial or helical arrays of microtubules in the vegetative cytoplasm of pollen tubes, especially in the cortical cytoplasm (fig. 3b; references in : Tiezzi, 1991; Pierson and Cresti, 1992). Microtubules persist in the oldest part of the pollen tube, and thus, the considerable reserve of tubulin that they represent is obviously not regenerated (Heslop-Harrison and Heslop-Harrison, 1988). Surprisingly, antitubulin hardly stains the tip region of the tube. The only reported exceptions are very short, just emerging pollen tubes where a bright fluorescence is observed in the very tip, which may indicate the presence of a pool of unpolymerized tubulin (Derksen et al., 1985). The scarcity of microtubules in the growth zone of pollen tubes together with the lack of a clear effect of colchicine treatment on pollen tube growth (review: Pierson and Cresti, 1992), raised the question of whether microtubules are actually necessary for tip growth in pollen tubes (see also Cresti and Tiezzi, this volume). At first sight the answer that they are not. However, antikinesin labeling with the monoclonal antibody K71S23 has revealed a punctuate pattern located in the cortex along the apical plasma membrane (fig. 3b; Moscatelli et al., 1988; Cai et al., 1992; Tiezzi et al., 1992). On account of this finding, Tiezzi and co-workers (accepted) suggested that, perhaps, a certain group of vesicles are transported to the tip of the pollen tube by a kinesin-microtubule locomotor system.

There are structural indications that microtubules may be involved in the orientation of cellulose microfibrils, but the concept is seriously weakened by discordant observations on root hairs (see Seagull, 1991). Pollen tubes of tobacco can be taken as an example both in favour of, and against this concept: random patterns of both microtubules and microfibrils are observed in the subapical zone (which does not obligatory mean co-alignment), and parallel alignment of microtubules together with criss-cross orientation of microfibrils are found in the other part of the pollen tube (Knuiman and Kroh, personal communication).

Microtubules sometimes appear to be interconnected (Franke et al., 1972). Co-localization of single microtubules or bundles of microtubules with microfilaments / actin filaments seem to be a current feature (Raudaskoski et al., 1987; Tiwari and Polito, 1988b; Lancelle and Hepler, 1989; Pierson et al., 1989). Associations between microtubules and tubular ER, microtubules and vesicles have been demonstrated by electron microscopy using both chemical fixation and cryo methods. One attractive consensus is that microtubules contribute to the positioning of the organelles in the cytoplasm and that they provide support for the more dynamic actin filaments that, in cooperation with myosin, assume the movement of the organelles. In animal systems, MAPs fulfil an important role in the regulation of microtubule-dependent processes. It is therefore of great interest to determine if MAPs are present in pollen tubes and if they are involved in the regulation of interactions between microtubules and organelles.

Research on more than twenty five species has revealed the occurrence of a considerable number of actin filaments in the pollen tube (fig. 2ab and 3b; reviewed in Pierson and Cresti, 1992). To judge from RF-FS preparations and some rhodamine-phalloidin preparations the arrays appear

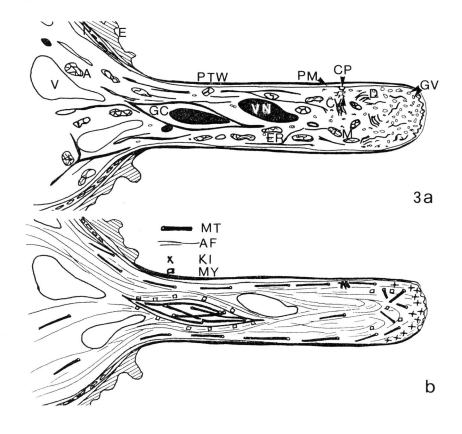

3a

b

Fig. 3. Schematic representation of the cytoplasmic components (a) and the cytoskeletal elements (b) that may occur in a pollen tube (longitudinal section).

a). Distinction can be made between the pollen tube wall (PTW), and the intine (I) and the exine (E) layer of the pollen grain wall, the plasma membrane (PM), vacuoles (V), the vegetative nucleus (VN), the spindle-shaped generative cell (GC) and numerous organelles, like Golgi vesicles (GV) in the tip region, dictyosomes (D) in the subapical zone , mitochondria (M), cisternae of endoplasmic reticulum (ER), amyloplasts (A), coated vesicles (CV) and coated pits (CP).

b). Microtubules (MT) are principally present in the cortical cytoplasm in net-axial orientation and in criss-cross like patterns in the subapical zone. They form thick bundles in the generative cell and converge in the tail-like extentions. Actin filaments (AF) can be observed in the entire vegetative cytoplasm. They may be more prominent around the generative cell and the vegetative nucleus. Myosin (MY) is associated with the surface of organelles and the surface of the generative cell. A kinesin-like protein (KI) appears to be present in small aggregates located in the tip region of the pollen tube. Other components presumed to be present in the pollen tube are profilin and clathrin, in association with coated pits.

to be very fine. A few rhodamine-phalloidin studies demonstrate the presence of actin filaments in the extreme tip (e.g. Pierson, 1988, Heslop-Harrison and Heslop-Harrison, 1991), but these observations are not supported by all data from RF-FS (Hepler, personal communication). In general, the orientation of the actin filaments is net-axial in the main part of the pollen tube and often wave-like in the subapical zone (fig. 3b).

The distribution of actin filaments coincides with the general pattern of cytoplasmic streaming, characterized as a "fountain streaming pattern" (e.g. Heslop-Harrison and Heslop-Harrison, 1987; Pierson et al., 1990). Tracking of the movement of single organelles has underlined the individuality of movement of each organelle (Heslop-Harrison and Heslop-Harrison, 1987;Pierson et al., 1990) and the diversity in behavior among the various types of organelles (Pierson et al., 1990). Golgi vesicles in the tip region show an intensive Brownian-like agitation, but mitochondria and amyloplasts show nearly smooth vectorial movement in the youngest part of the tube and saltatory movements in the thin layer of cytoplasm bordering the large vacuoles. Putative endoplasmic reticulum tubules display continuous undulatory movements and changes in configuration (Pierson et al., 1990; for schematic representations of the location of organelles in pollen tubes see fig. 3a).

For a long time it has been known that cytochalasins are strong inhibitors of organelle movement in pollen tubes (e.g. Franke et al., 1972) and that actin is present in pollen (Condeelis, 1974). The postulation that the movement of organelles in pollen tubes is generated by acto-myosin-ATP interactions (Condeelis, 1974) has encountered much support from later experiments. Cytochalasins also appear to stop the transport of Golgi vesicles to the tip (Picton and Steer, 1981). Antibodies directed against different subunits of myosin show a positive reaction for the surface of organelles (fig. 3b; Tang et al., 1989a; Heslop-Harrison and Heslop-Harrison, 1989a and b). Kohno and Shimmen (1988) elegantly demonstrated that organelles of pollen tubes of *Lilium longiflorum* can perform movement along Chara actin filament bundles when supplied with ATP and Mg^{2+}. This movement is slowed by treatment with NEM and p-CMB, typical inhibitors of myosin dependent movement, and a concentration of vanadate exceeding a 100 micromolar concentration. Moreover, constructed fluorescent actin filaments begin to move when a crude extract of pollen tubes is added (Kohno et al., 1990). In pollen tubes of *Tradescantia,* a CDPK has been found which co-localize with actin filaments and which inactivates a myosin-like peptide in a Ca^{2+} dependent manner (Putnam-Evans et al., 1989). A monoclonal antibody against this CDPK did not react with purified actin from rabbit muscle or Dictyostelium and did not bind to proteins having a similar molecular weight to actin (Putnam-Evans et al., 1989).

In conclusion, the existence of an actomyosin driven system in pollen tubes is highly probable. It is, however, not clarified whether all organelles move according to this mechanism onle. We have already mentioned results on the presence of a kinesin-like protein in pollen tubes and the concept that a microtubule-kinesin locomotor apparatus might be utilized for the transport of vesicles to the tip region. Do Golgi vesicles within the extreme tip move along actin filaments, along microtubules or are they unbound particles fusing at random with the plasma membrane? We have no answer yet, and we can assume that the reality is probably rather complex and that it includes fine regulatory mechanisms triggered by secondary factors, for example the local pH and local concentrations of calcium ions, respectively calcium sequestering molecules such as calmodulin and CDPK (Tirlapur, this volume).

Besides the aspect of intracellular movement Steer (1990) hypothetized that a meshwork of actin filaments in the extreme tip of the tube would prevent the plasma membrane from being disrupted by osmotic pressure. This actin meshwork would be loose enough to permit extension of the tip. It is possible that cross-linking proteins, such as ankyrin and spectrin, cooperate together with actin filaments to constitute an "elastic scaffolding" in the pollen tube tip. Calcium could also play a role in modulating the rigidity of the cytoskeleton-membrane complex (Tirlapur, this volume).

The last cytoskeletal protein that has been discovered in pollen is profilin (Valenta et al., 1991). Profilins are typical sequestering molecules; they have the common property of forming a 1:1 complex with G-actin (Kd about 10^{-5} M), which seriously diminishes the ability of the actin monomer to polymerize (Bershadsky and Vasiliev, 1988). Apparently, profilins do not interact with F-actin. The role of profilin in pollen is still unclear, and we can only guess that the protein may participate to the regulation of the amount of activelyavailable actin in the cell.

The generative cell and the sperm cell cytoskeleton

The reports dealing with the cytoskeleton of the generative cell and the sperm cell of Angiosperms agree in showing a prominent system of microtubule bundles aligned obliquely or parallel to the long axis of the cell (review: Palevitz and Tiezzi, 1992; fig. 1a and 3b; see also fig. 3c in Pierson and Rennoch, this volume). Immunofluorescence microscopy reveals numerous branching and convergence of the bundles, which together form a basket-like structure, and electron dense cross-bridges between microtubules (in review: Tiezzi, 1991). The microtubular system of angiosperms generative/sperm cells shows morphological similarities with the axoneme of flagelled sperms of lower plants and animals (Tiezzi, 1991). However, there is no cytological evidence -yet- that in higher plants the locomotion of the male gametes is also an autonomous process, driven by forces coming from interaction between dynein and microtubules. In contrast, myosin epitopes (Tang et al., 1989a; Heslop-Harrison and Heslop-Harrison, 1989a and 1989b) and strip-like projections (Cresti et al., 1991) have been shown on the surface of the vegetative membrane of the generative cell. The spatial organization of actin filaments around the generative cell (fig. 2d; reviewed in Palevitz and Tiezzi, 1992) and the effect of cytoskeleton inhibitors (colchicine, nocodazole, cytochalasin) tend to suggest that actin filaments of the vegetative cytoplasm together with a myosin coating on the vegetative face of the outer membrane of the generative cell are responsible for the movement of these cells.

The overwhelming majority of data indicates that there are no actin filaments inside the generative cell, either in interphase or during division (Palevitz and Tiezzi, 1992). The only report in which actin filaments have been explicitly demonstrated in the sperm cell (Taylor et al., 1989) is now under reinvestigation.

To the best of our knowledge, no arrays of microtubules arranged like those of the preprophase band of somatic cells have been reported from dividing generative cells. In some species, e.g. *Nicotiana tabacum* (Raudaskoski et al., 1987), a regular metaphase plate perpendicular to the cell axis is observed, while in other species, e.g. *Tradescantia virginiana* (Liu and Palevitz, 1991 and references herein) the plane of division is clearly oblique. In the latter case, the kinetochores visualized by using the serum of CREST patients, are disposed in a tandem fashion at metaphase (Liu and Palevitz, 1991). Two types of mechanisms have been distinguished to be effective at the end of karyokinesis and during cytokinesis: the first, occurring for instance in *Tradescantia virginiana*, is characterized by a furrowing or constriction process, the second, as in *Nicotiana tabacum*, includes the formation of a distinct microtubule phragmoplast and a cell plate (reviewed by Palevitz and Cresti, 1992). In *Allemanda neriifolia*, the rearrangement of the microtubules was followed at various stages of the second haploid division after isolation of the generative cell (Zee and Aziz-Un-Nisa, 1991)

Final remark

While the data presented in this mini-review are intended to provide an overview, they also evidence the numerous questions which remain to be elucidated, as a challange for the students to whom this volume is dedicated...

Acknowledgements

This work was supported by grants from the European Communities (grants BAP-0597-I-CH and BIOT-0078-I-CH to ESP) and the Ministry of Foreign Affairs of Italy (grant to LYQ), in the framework of cultural and scientific exchanges between Italy and the People's Republic of China. The authors thank Dr. V. Polito and Dr. S.Tiwari, Springer-Verlag and the Wissenschaftliche Verlagsgesellschaft Stutgart for giving permission to use published results as a source for figures 1 and 2. They also want to acknowledge the various colleagues who kindly provided preprints of their work and Mr. G. Fabbri for making the drawings.

References

Åström H, Virtanen I, Raudaskoski M (1991) Cold-stability in the pollen tube cytoskeleton. Protoplasma 160: 99-107

Bershadsky AD, Vasiliev JM (1988) Cytoskeleton. Plenum Press New York London

Cai G, Bartalesi A, Moscatelli A, Del Casino C, Tiezzi A, Cresti M (1992) Microtubular motors in the pollen tube of *Nicotiana tabacum*. Proc Int Symp on Angiosperm Pollen and Ovules: Basic and applied aspects Villa Olmo, Como June 23-27, 1991 (in press)

Condeelis JS (1974) The identification of Factin in the pollen tube and protoplast of Amaryllis belladonna. Exptl Cell Res 88: 435-439

Cresti M, Pacini E, Ciampolini F, Sarfatti G (1977) Germination and earty tube development "in vitro" of *Lycopersicon peruvianum* pollen: ultrastructural features. Planta 136: 239-247

Cresti M, Hepler PK, Tiezzi A, Ciampolini F (1986) Fibrillar structures in Nicotiana pollen: changes in ultrastructure during pollen activation and tube emission. In: Biotechnology and ecology of pollen. Mulcahy DL, Mulcahy GB, Ottaviano E (eds) Springer Verlag Berlin Heidelberg New York pp. 283-288

Cresti M, Ciampolini F, Van Went JL (1991) Strip-shaped projections at the cytoplasmic face of the outer membrane of the generative cell in *Amaryllis belladonna*. Ann Bot 68: 105-107

Derksen J, Pierson ES, Traas JA (1985) Microtubules in vegetative and generative cells of pollen tubes. Eur J Cell Biol 38: 142-148

Derksen J, Emons AM (1990) Microtubules in tip growth systems. In: The growth in plant and fungal cells. Heath IB (ed) Academic Press San Diego New York Boston London Sydney Tokyo Toronto pp.147-181

Dustin P (1984) Microtubules. Springer Verlag Berlin New York

Emons AMC, Pierson E, Derksen J (1991) Cytoskeleton and intracellular movement in plant cells. In: Biotechnology: current progress. Cheremisinoff PN, Ferrante LM (eds) Technomic Publishing to Lancaster Basel pp. 311-335

Fosket DE (1989) Cytoskeletal proteins and their genes in higher plants. In: The biochemistry of plants; a comprehensive treatise. Marcus A (ed) Vol. 15.

Molecular Biology. Academic Press San Diego New York Berkeley Boston London Sydney Tokyo Toronto. pp. 393-455

Franke WW, Herth W, Van Der Woude WJ, Morré DJ (1972) Tubular and filamentous structures in pollen tubes: possible involvement as guide elements in protoplamic streaming and vectorial migration of secretory vesicles. Planta 105: 317-341

Goldstein LSB (1991) The kinesin superfamily: tails of functional redundancy. Trends Cell Biol 1: 93-98

Hammer JA III (1991) Novel myosins. Trends Cell Biol 1: 50-56

Heslop-Harrison J, Heslop-Harrison Y, Cresti M, Tiezzi A, Ciampolini F (1986) Actin during pollen germination. J Cell Sci 86: 1-8

Heslop-Harrison J, Heslop-Harrison Y (1987) An analysis of gamete and organelle movement in the pollen tube of *Secale cereale* L. Plant Sci 51: 203-213

Heslop-Harrison J, Heslop-Harrison Y (1988) Site of origin of the peripheral microtubule system of the vegetative cell of the angiosperm pollen tube. Ann Bot 62: 455-461

Heslop-Harrison J, Heslop-Harrison Y (1989a) Myosin associated with the surfaces of organelles, vegetative nuclei and generative cells in angiosperm pollen grains and tubes. J Cell Sci 94: 319-325

Heslop-Harrison J, Heslop-Harrison Y (1989b) Actomyosin and movement in the angiosperm pollen tube: an interpretation of some recent results. Sex Plant Reprod 2: 199-207

Heslop-Harrison J, Heslop-Harrison Y (1991) The actin cytoskeleton in unfixed pollen tubes following microwave-accelerated DMSO- permeabilisation and TRITC-phalloidin staining. Sex Plant Reprod 4: 6-11

Kohno T, Shimmen T (1988) Accelerated sliding of pollen tube organelles along *Characeae* actin bundles regulated by Ca^{2+}. J Cell Biol 106: 1539-1543

Kohno T, Chaen S, Shimmen T (1990) Characterization of the translocator associated with pollen tube organelles. Protoplasma 154: 179-183

Kohno T, Okagaki T, Kohama K, Shimmen T (1991) Pollen extract supports the movement of actin filaments in vitro. Protoplasma 161: 75-77

Kristen U (1986) General and molecular cytology: the cytoskeleton: microtubules. Progress in Botany Vol 48 Springer Verlag Berlin Heidelberg

Kristen U (1987) General and molecular cytology: the cytoskeleton: microfilaments. Progress in Botany Vol 49 Springer Verlag Berlin Heidelberg

Lancelle SA, Callaham DA, Hepler PK (1986) A method for rapid fixation of plant cells. Protoplasma 131: 153-165

Lancelle SA, Hepler PK (1989) Immunogold labelling of actin on sections of freeze-substituted plant cells. Protoplasma 150: 72-74

Liu B, Palevitz BA (1991) Kinetochore fiber formation in dividing generative cells of *Tradescantia*: Kinetochore reorientation associated with the transition between lateral microtubule interactions and end-on kinetochore fibers. J Cell Sci 98: 475-482

Ludwig SR, Oppenheimer DG, Silflow CD, Snustad DP (1988) The alfa-tubulin gene ot *Arabidopsis thaliana*: primary structure and preferential expression in flowers. Plant Mol Biol 10: 311-321

Moscatelli A, Tiezzi A, Vignani R, Cai G, Bartalesi A, Cresti M (1988) Presence of kinesin in tobacco pollen tube. In: Sexual reproduction in higher plants. Cresti M, Pacini E, Gori P, (eds) Springer Verlag Berlin Heidelberg New York London Paris pp. 205-209

Palevitz BA, Tiezzi A (1992) The organization, composition and function of the generative cell. Int Rev Cytol (accepted)

Perdue TD, Parthasarathy MV (1985) In situ localization of F-actin in pollen tubes. Eur J Cell Biol 39: 13-20

Picton JM, Steer MW (1981) Determination of secretory vesicle production rates by dictyosomes in pollen tubes of *Tradescantia* using cytochalanin D. J Cell Sci 49: 261-272

Pierson ES (1988) Rhodamine-phalloidin staining of actin in pollen after dimethyl sulphoxide permeabilization: a comparison with the conventional formaldehyde preparation. Sex Plant Reprod 1: 83-87

Pierson ES, Derksen J, Traas JA (1986) Organization of microfilaments and microtubules in pollen tubes grown in vitro or in vivo in various angiosperms. Eur J Cell Biol 41: 14-18

Pierson ES, Kengen HMP, Derksen J (1989) Microtubules and actin filaments co-localize in pollen tubes of *Nicotiana tabacum* L. and *Lilium longiflorum* Thunb. Protoplasma 150: 75-77

Pierson ES, Lichtscheidl IK, Derksen J (1990) Structure and behaviour of organelles in living pollen tubes of *Lilium longiflorum*. J Exptl Bot 41: 1461-1468

Pierson ES, Cresti M (1992) Cytoskeleton and cytoplasmic organization of pollen and pollen tubes. Int Rev Cytol (accepted)

Putnam-Evans C, Harmon AC, Palevitz BA, Fechheimer M, Cormier MJ (1989) Calcium-dependent protein kinase is localized with F-actin in plant cells. Cell Motility Cytoskeleton 12: 12-22

Raudaskoski M, Åström H, Perttilä K, Virtanen I, Louhelainen J (1987) Role of the microtubule cytoskeleton in pollen tubes: an immunocytochemical and ultrastructural approach. Biol Cell 61: 177-188

Seagull RW (1991) Role of the cytoskeletal elements in organized wall microfibril deposition. In: Biosythesis and biodegretation of cellulose and cellulosic materials. Haigler CH, Weimer P (eds). Marcel Dekker Press

Steer MW (1990) Role of actin in tip growth. In: The growth in plant and fungal cells. Heath IB (ed) Academic Press San Diego New York Boston London Sydney Tokyo Toronto pp.

Stinson JR, Eisenberg AJ, Willing RP, Pè ME, Hanson DD, Mascarenhas JP (1987) Genes expressed in the mole gametophyte of flowering plants and their isolation. Plant Physiol 83: 442-447

Tang XJ, Hepler PK, Scordilis SP (1989a) Immunochemical and immunocytochemical identification of a myosin heavy chain polypeptide in *Nicotiana* pollen tubes. J Cell Sci 92: 569-574

Tang XJ, Lancelle SA, Hepler PK (1989b) Fluorescence microscopic localization of actin in pollen tubes: comparison of actin antibody and phalloidin staining. Cell Motility Cytoskeleton 12: 216-224

Taylor P, Kenrick J, Li Y, Kaul V, Gunning BES, Knox RB (1989) The male germ unit of *Rhododendron*: quantitative cytology, three-dimensional reconstruction, isolation and detection using fluorescent probes. Sex Plant Reprod 2: 254-264

Tewinkel M, Kruse S, Quader H, Volkmann D, Stevens A (1989) Visualization of actin filament pattern in plant cells without pre-fixation: a comparison of differently modified phallaloxins. Protoplasma 149: 178-182

Tiezzi A (1991) The pollen tube cytoskeleton. Electron Microsc Rev 4: 205-219

Tiezzi A, Cresti M (1990) The cytoskeleton during pollen tube growth and sperm cell formation. In: Mechanism of fertilization: plant to human. Dale B (ed) NATO ASI Vol H45 Springer Verlag Berlin Heidelberg pp. 17-34

Tiezzi A, Moscatelli A, Cai G, Bartalesi A, Cresti M (1992) An immunoreactive homolog of mammalian kinesin in *Nicotiana tabacum* pollen tubes. Cell Motility Cytoskeleton 21: 132-137

Tiwari SC, Polito VS (1988a) Spatial and temporal organization of actin during hydration, activation, and germination of pollen in *Pyrus communis* L.: a population study. Protoplasma 147: 5-15

Tiwari SC, Polito VS (1988b) Organization of the cytoskeleton in pollen tubes of *Pyrus communis*: a study employing conventional and freeze-substitution electron microscopy, immunofluorescence and rhodamine phalloidin. Protoplasma 147: 100-112

Tiwari SC, Polito VS (1990) The initiation and organization of microtubules in germinating pear (*Pyrus communis* L.) pollen. Eur J Cell Biol 53: 384-389

Tiwari SC, Polito VS, Webster BD (1990) In dry pear (*Pyrus communis* L.) pollen, membranes assume a tightly packed multilamellate aspect that disappears rapidly upon hydration. Protoplasma 153: 157-168

Valenta R, Duchêne M, Pettenburger K, Sillaber C, Valent P, Bettelheim P, Breitenbach M, Rumpold H, Kraft D, Scheiner O (1991). Identification of profilin as a novel pollen allergen; IgE autoreactivity in sensitised individuals. Science 253: 557-560

Warner FD, McIntosh JR (eds) (1989) Cell movement: kinesin, dynein, and microtubule dynamics Vol 2. Alan R Liss New York

Warner FD, Satir P, Gibbons IR (eds) (1989) Cell movement: the dynein ATPases Vol 1. Alan R Liss New York

Yen LF, Liu X, Liu GQ (1992) In vitro polymerization of F-actin and assembly of microfilament cytoskeleton of pollen actin. Proc Int Symp on Angiosperm Pollen and Ovules: Basic and applied aspects Villa Olmo, Como June 23-27, 1991 (in press)

Zee SY, Aziz-Un-Nisa (1991) Mitosis and microtubule organizatorial changes in isolated generative cells of *Allemanda neriifolia*. Sex Plant Reprod 4: 132-137

SPOROPHYTIC AND GAMETOPHYTIC SELF-INCOMPATIBILITY

H. Kaufmann, H. Kirch, T. Wemmer, A. Peil, F. Lottspeich, H. Uhrig, F. Salamini and R. Thompson
Max-Planck-Institut für Züchtungsforschung, Carl von Linné-Weg 5000 Köln 30, Germany

Sporophytic and gametophytic self-incompatibility

In this article the main features of the best-studied self- incompatibility (SI) systems will be summarized (see also Nasrallah and Nasrallah 1989, and Harings et al. 1991), followed by a more detailed description of potato pistil proteins and their possible role in SI.

Self-incompatibility is a mechanism for preventing self- fertilization which is widespread among flowering plants. The ability of a pollen grain to fertilise depends either on its own genotype at the S-locus (gametophytic SI) or on the S-genotype of the pollen parent (sporophytic SI). Fertilization is only possible when the S-allele phenotype of the pollen grain differs from the S-alleles expressed in the pistil. The S-gene products expressed in pollen and pistil are thought to interact and mediate this recognition process, but this interaction has yet to be demonstrated at the molecular level.

Over the past ten years, a number of S-linked pistil proteins have been identified, including examples in Brassica and Solanaceous species. More recently, the corresponding genes encoding Brassica SLGs (S-linked glycoproteins) and S-associated polypeptides from *Nicotiana alata, Solanum tuberosum, Petunia hybrida* and *Petunia inflata*, have been isolated. In Table 1 the main features of the S-associated proteins from both SI systems are compared.

Both protein types are N-glycosylated and are synthesized bearing a signal peptide which is necessary for their extracellular localization. When the sequences of several alleles are compared, a number of highly conserved regions, including several cysteine residues, are apparent. These are thought to be involved in the formation of a "core" of conserved secondary structure. A number of highly variable stretches are also present, which contain the

Table 1:

Comparison of S-associated glycoproteins from gametophytic (Solanaceae) and sporophytic (Brassicaceae) species

	Solanaceae	Brassicaceae
Mr	24 - 30 kd	45 - 50 kd
pI	8 - 10	8 - 10
N-Glycosylation	+	+
Interallelic homology	50 - 70 %	70 - 90 %
Associated activity	Ribonuclease	Matrix-forming protein ?
Localization	Transmitting tissue	Stigmatic papillae

putative N-glycosylation sites. These hypervariable regions are thought to be responsible for the allele-specific surface structure for each S-allele protein.

Despite a number of common features, the Brassica and Solanaceae S-linked proteins lack sequence homology with one another, and probably operate via different mechanisms. The Solanaceae S- linked proteins are also referred to as S-RNAses, as they possess ribonuclease activity, a property not shared by the Brassica SLGs. These due to sequence homology with collagen VI have been proposed to play a role in pollen water economy.

Stein et al. (1991) have recently identified a kinase gene, SRK, which is tightly linked and related in sequence to the Brassica SLG gene. SRK transcripts were detected in both pistil and anther. The predicted protein consists of an SLG-related N- terminal, connected by a putative membrane-spanning domain to a C-terminal sequence homologous to animal tyrosine kinases. It has been suggested that SI may be controlled by the presence of the SRK gene product or by an interaction between SRK and SLG proteins. One would, however, expect a pollen membrane-localized SRK to be encoded by the male gametophyte itself - this is hard to reconcile with the sporophytic inheritance of SI in Brassica. If the stigma-localized SRK were to be involved in determining SI, the nature of the pollen ligand would still be unclear.

In order to examine the possible intracellular effect of S-RNAse action, McClure et al. (1990) labelled pollen with 32P-PO4, and used it in a series of incompatible and compatible pollinations on unlabelled Nicotiana styles. RNA was subsequently extracted and analysed for integrity. Pollen RNA in incompatible pollinations was significantly more degraded than in compatible pollinations, and degraded RNA was restricted to the tip-proximal part of the pollen tube i.e., where the SI reaction is thought to occur.

How is the selective degradation of RNA from incompatible pollen tubes achieved? As it depends on the S-genotype of the pollen grain, this is presumably via an S-locus-derived gene product. The classical model of S-gene structure predicts that one structural gene ("Specificity segment") is expressed from both style and pollen promoters ("Activity segments"). However, the S- RNAse sequence does not seem to be expressed in pollen, and therefore at least for the Solanaceae, it appears likely that the pollen-S-gene product is encoded by another sequence in the vicinity of the S-RNAse gene.

Brassica SLG genes and promoter/reporter constructs have been introduced into transgenic tobacco, Arabidopsis and oilseed rape plants. In general, their expression pattern in Brassica species reflects that predicted from the protein distribution in untransformed Brassica plants, i.e., expression in the pistil is restricted to the stigmatic papillae, whereas in tobacco (N. tabacum), Brassica SLG promoter/reporter constructs were expressed in the transmitting tract and the placental epidermis surrounding the ovules, i.e., the sites of expression of Solanaceous S-RNAses. Presumably, the cell-type-specific factors which direct S-gene promoters are conserved but are restricted to the cells in which the self-incompatibility phenotype is expressed in each system. More recent data in which B. oleracea plants were transformed with SLG promoter/GUS constructs produced expression in tapetum and pollen cells, contrary to the prediction of a strictly sporophytic control (Sato et al. 1990). It appears that an element conferring gametophytic expression is present in this Brassica SLG promoter fragment. Possibly the evolution of the sporophytic control from gametophytic control occurred via recruitment of a cis-repressor of this gametophytic element, and this repressor is absent or ineffective in the transgenic plants tested.

118

Using sensitive reporter gene constructs, evidence has been obtained for the expression of a Brassica SLG-promoter in anther tissue. In transgenic Arabidopsis, the promoter was expressed in the (sporophytic) tapetum, whereas in *N. tabacum*, expression was detected in pollen grains. These results are consistent with the observed S-phenotypes observed in both systems: in Brassica, the S-locus gene product is putatively encoded by the tapetum, and modifies the maturing pollen grains externally (e.g. by deposition in the exine coat). In SI Solanaceous species, the S- locus gene product is encoded by the male gametophyte and is cell-autonomously located.

The work of the Cologne group began with the analysis of pistil proteins present in diploid potato lines of known S-genotypes (Fig. 1). An abundant, basic glycoprotein could be associated to each of the S-alleles S1 through S4 present in the material (Kirch et al. 1989.) In addition, two further basic proteins, one glycosylated (SK1) and one unglycosylated (SK2) were present irrespective of genotype. The S-linked proteins S2-S4 were purified and sequenced, and found homologous to S-RNAses from *Nicotiana alata* and *Lycopersicon peruvianum*. cDNA and genomic clones were isolated for the S1

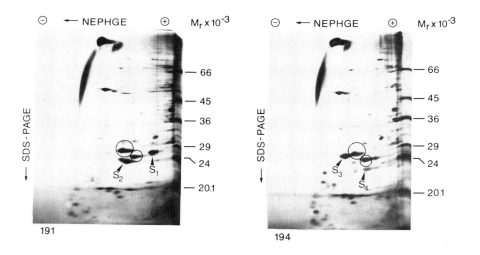

Figure 1. Two-dimensional gel electrophoresis of diploid potato clones 191 (S1S2) and 194 (S3S4). Total pistil proteins were fractionated and detected by silver staining. Each extract contains S-allele-specific proteins (arrowed), as well as two constant polypeptides, encircled. Large circle = SK1, small circle = SK2 (from Kirch et al. 1989).

and S2 alleles, and cDNA or PCR probes have been subsequently obtained for several other alleles. The sequences of these alleles vary greatly in their interallelic homology (Fig. 2). We have defined four classes of S-alleles which do not cross-hybridize between classes (defined by S1, S2, S4 and Sc) to which most potato S-alleles so far tested belong. Within these classes, cross-hybridizing alleles such as S1 and S3 (in the S1 class) appear nevertheless to be fully co-dominant in the style. The two most similar alleles so far sequenced differ by only 5% of their residues. 13/27 of these differences are located in one hypervariable region adjacent to the single intron present in the gene (Kaufmann et al. 1991).

When S-RNAse sequences from different Solanaceous species are compared, inter-species similarities are often greater than intra-species similarities for allele-pairs. This indicates that the proliferation of alleles at the S-locus preceded species divergence, a point developed in more detail by Kao and colleagues (Ioerger et al. 1990). Evidence that the hypervariable residues are located on the surface of S-RNAses is provided by the reaction of S-RNAses from potato alleles S1 through S4 with antiserum raised to S-RNAse-S2. Only S-RNAse-S2 reacts with the antiserum. Similarly, antiserum raised against S-RNAse-S1 reacts only with S-RNAses-S1 and S3, and not with S2 or S4 (Fig. 3). Despite surface structural dissimilarities, all potato S-RNAses so far examined possess ribonuclease activity. This has been determined either by incubation of purified protein fractions eluted from a mono-S column with labelled RNA (Fig. 4) or by overlaying IEF-separated proteins with an RNA-containing agarose gel, which after incubation is stained with ethidium bromide. Using the latter method it was apparent that in contrast to the S-associated proteins, neither SK1 nor SK2 had RNAse activity. As this activity is conserved, it is presumably involved in the mechanism of action of SI in Solanaceae, as mentioned earlier, and the mediation of RNA degradation one would predict must involve the pollen S-gene product. We have been unable to detect a pollen transcript or protein homologous to the S-RNAses, and are therefore looking for a discrete pollen ORF in the genomic neighbourhood of the S-locus.

The mechanism of action of the S-locus may be better understood with the aid of self-compatible (SC) mutants. We have studied two types of mutant, first, an SC mutant which was believed to correspond to a translocation of the S1-allele (Thompson et al. 1991).
This mutation was effective in pollen but had no phenotype on the stylar side

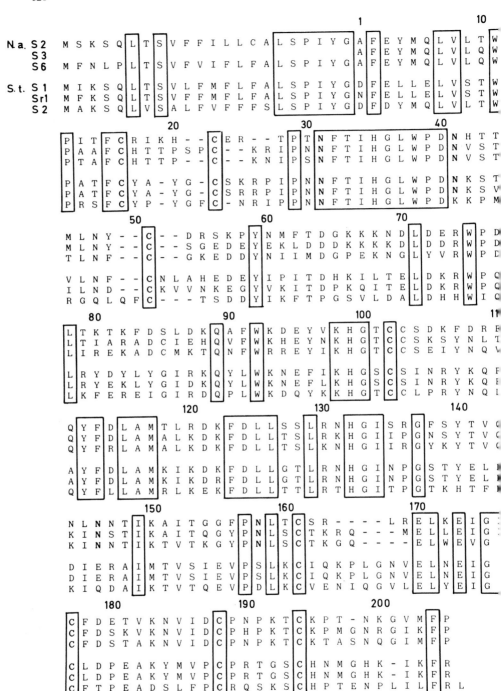

2A

	StS_{r1}	StS_1	StS_2	NaS_2	NaS_3	NaS_6	PiS_1	PiS_2	PiS_3	PhS_1	PhS_2	PhS_3
StS_{r1}	–											
StS_1	95	–										
StS_2	66	66	–									
NaS_2	64	66	63	–								
NaS_3	61	61	58	75	–							
NaS_6	61	62	62	75	80	–						
PiS_1	64	68	77	59	59	58	–					
PiS_2	67	67	77	61	56	56	84	–				
PiS_3	67	68	79	62	55	58	84	89	–			
PhS_1	67	65	58	61	60	59	60	59	58	–		
PhS_2	66	66	77	58	55	56	84	95	89	60	–	
PhS_3	63	64	74	51	51	54	82	89	87	55	89	–

Figure 2. A) Comparison of deduced proteins sequences of S- RNAses from three *Nicotiana alata* (N.a.) and three *Solanum tuberosum* (S.t.) S-alleles. Residues conserved between all six alleles are boxed. The intron interrupts the coding sequence between codons 64 and 65. Histidine residues 35 and 99 are thought to form part of the ribonuclease active site.

B) Percent amino acid sequence similarity between S-RNAses as determined by UWGCG Bestfit (Devreux et al. 1984, Nucl Acids Res 12: 387-395).

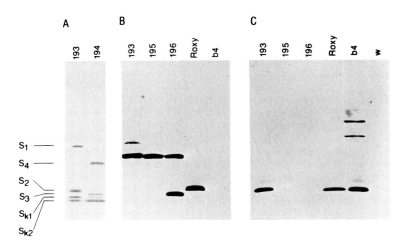

Figure 3. Immunoblot analysis of IEF-gel fractionated potato pistil proteins. A) Silver-stained pistil extracts from potato clones 193/2 (S1S2) and 194/1 (S3S4). B) Incubation with S1-RNAse antiserum. C) Incubation with S2-RNAse antiserum. Genotypes 193 (S1S2), 195 (S1S4), 196 (S1S3), Roxy (SR1, SR2), b4 and w, two *Solanum berthaultii* lines of unknown S-allele constitution.

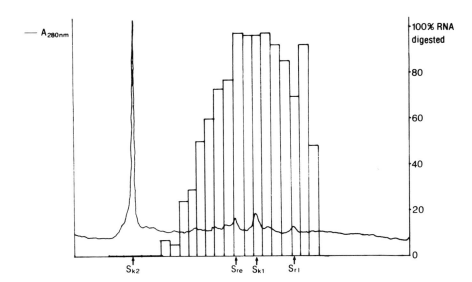

Figure 4. Assay of Ribonuclease activity in a potato pistil extract. 1 mg basic proteins purified from 2 g mature potato pistils of cultivar "Roxy" was applied to a 10 ml Pharmacia Mono-S FPLC column and fractionated with a 50 + 50 ml 0-0.5 M NaCl gradient. RNAse activity was determined in each fraction by incubation with 32P-RNA and assessment of % loss of TCA precipitability. Although SK1 polypeptide co-eluted with RNAse activity in this assay, a second RNAse-assay, in which IEF-separated proteins were stained in situ, showed this protein has no RNAse activity.

(Fig. 5). The cellular and subcellular localization of SK2 in potato styles was determined using immuno cytochemistry. At the light microscope level, SK2 was restricted to the conducting tissue (transmitting tract), and did not extend to the stigma surface. The placental epidermis in the ovary also reacted weakly with SK2 antibody. At the E/M level, the protein was restricted to the intercellular matrix. Thus, the distribution of SK2 mirrors exactly that of the S-RNAses.

What is the function of the SK2-chitinase? A number of pathogenesis-related protein mRNAs have been detected in the inflorescence including chitinases, a glucanases, and proteinase inhibitors. The location of SK2 would be consistent with a role in protecting the ovary against pollen-transmitted fungal or bacterial ingress. Its restricted distribution suggests it may possibly only be of selective advantage in the presence of certain species-specific pathogens.

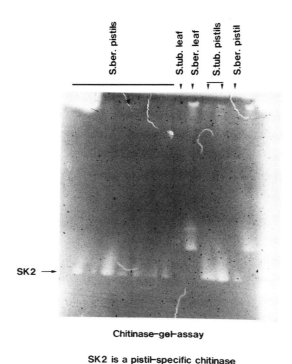

Chitinase–gel–assay

SK2 is a pistil–specific chitinase

Figure 5. Chitinase gel assay. Leaf and pistil extracts from *Solanum tuberosum* and *Solanum berthaultii* were fractionated on a non-denaturing IEF gel and electroblotted onto nitrocellulase. The filter was overlaid with a substrate gel containing Remazol Brilliant violet-derivatized chitin, and protein was blotted from filter to substrate gel where clearing of the violet dye indicates chitinase activity (G Jach, J Logemann, MPI f. Zchtungsforschung, Pers. commun.).

The interaction between germinating pollen and pistil is the site of a number of specific biochemical interactions including those mediating self-incompatibility. Attempts to demonstrate binding of S-RNAses or SK2 to particular pollen components have, however, in our hands been so far unsuccessful. The mechanism by which S- RNAses selectively act on pollen tubes according to their S- allele constitution remains a priority for this and other groups in the field.

It conferred self-compatibility on pollen grains carrying S-alleles other than S1, hence the suggestion that it might consist of a translocated S1-allele. However, we could find no evidence for S1-RNAse sequences associated with this mutation by hybridization. From this we conclude that the mutation may involve just a non-homologous 'pollen component' of the S1-allele or alternatively is in another non-S-gene which can show an S- allele-specific discrimination.

The second mutant phenotype was present in a self-compatible potato species, *Solanum verrucosum*. This species lacks S-RNAse polypeptides on IEF gel separations, which suggested the hypothesis that self-compatibility may be due to the absence of S-RNAses. An S1-allele was introduced from S. tuberosum via an interspecific cross into *S. verrucosum*. The S1-RNAse is synthesized at wild-type levels in the F1 hybrid. However, on pollination with S1S4 pollen (from self-incompatible *S. tuberosum*), amongst other genotypes, S1 homozygotes were obtained, i.e., the S1-allele in the F1-hybrid background did not confer incompatibility against S1-bearing pollen grains. Therefore, it can be concluded that *S. verrucosum* possesses a dominant style-expressed suppressor of S1-function. Further experiments to genetically characterize the suppressor are planned.

As genetic data suggest the possible involvement of other genes in mediating SI response, we have characterized a further abundant stylar protein, SK2 (Fig. 1). SK2 shows little allelic variation within potato in contrast to the S-RNAse locus. Three alleles have been so far identified, one of which is null. No allelic variation in SK2 can be associated with allelic variation at the S-locus. Antiserum was raised against FPLC-purified SK2, and used to study the distribution of the protein throughout the plant. SK2 is essentially restricted to the ovary and pistil by Western blot analysis, with a much lower level in primary roots. SK2 has a restricted distribution within the Solanaceae, being present in potato and tomato pistils, but absent from tobacco and petunia.

Protein sequence determination on SK2 was restricted to sequencing of internal peptides, as the N-terminal was blocked. Oligonucleotides were synthesized on the basis of two peptides and used to generate a PCR probe for cDNA and genomic clone isolation. The cDNA sequence was 70% homologous to a previously sequenced potato endochitinase gene (Laflamme and Roxby 1989) and FPLC-pure SK2 was shown to have chitinase activity

References

Haring V, Gray JE, McClure BA, Anderson MA, Clarke AE (1991) Self-incompatibility: A self-recognition system in plants. Science 250: 937-941

Ioerger TR, Clark AG, Kao TH (1990) Polymorphism at the self-incompatibility locus in Solanaceae predates speciation. Proc Natl Acad Sci USA 87: 9732-9735

Kaufmann H, Salamini F, Thompson RD (1991) Sequence variability and gene structure at the self-incompatibility locus of *Solanum tuberosum*. Mol Gen Genet 226: 457-466

Kirch HH, Uhrig H, Lottspeich F, Salamini F, Thompson RD (1989) Characterization of proteins associated with self- incompatibility in *Solanum tuberosum*. Theor Appl Genet 78: 581-588

Laflamme D, Roxby D (1989) Isolation and nucleotide sequence of cDNA clones encoding potato chitinase genes. Plant Mol Biol 13: 249-250

McClure BA, Gray JE, Anderson MA, Clarke AE (1990) Self-incompatibility in *Nicotiana alata* involved degradation of pollen rRNA. Nature 347: 757-760

Nasrallah ME, Nasrallah JB (1989) The molecular genetics of self-incompatibility in *Brassica*. Ann Rev Genet 23: 121-139

Sato T, Thorsness MK, Kandasamy MK, Nishio T, Hirai M, Nasrallah JB, Nasrallah ME (1991) Activity of an S- locus gene promoter in pistils and anthers of transgenic *Brassica*. Plant Cell 3: 867-876

Stein JC, Howlett B, Boyes DC, Nasrallah ME, Nasrallah JB (1991) Molecular cloning of a putative receptor protein kinase gene encoded at the self-incompatibility locus of *Brassica oleracea*. Proc Natl Acad Sci USA 88: 8816-8820

Thompson RD, Uhrig H, Hermsen JGTh, Salamini F, Kaufmann H (1991) Investigation of a self-incompatible mutation in *Solanum tuberosum* clones inhibiting S-allele activity in pollen differentially. Mol Gen Genet 226: 283-288

APPLICATION OF IN VITRO POLLINATION AND FERTILIZATION TECHNIQUES FOR BREEDING AND GENETIC MANIPULATION OF LILIUM

R.J. Bino, M.G.M. van Creij, L.M. van der Leede-Plegt, A.J. van Tunen and J.M. van Tuyl
DLO-Centre for Plant Breeding and Reproduction Research (CPRO-DLO)
P.O. Box 16, 6700 AA Wageningen
The Netherlands

Introduction

Application of in vitro pollination and fertilization for breeding

Possibilities for cross combinations in *Lilium* are limited by incompatibility and incongruity. Sexual barriers preventing interspecific hybridization have been separated into pre- and post- fertilization barriers. In lily, many studies have dealt with methods for overcoming the stylar pre-fertilization barriers (Ascher and Peloquin, 1966; Van Tuyl et al., 1982). In a comparison of several pollination methods, it was concluded that in lily pre- fertilization barriers can be bypassed using the cut-style technique (Van Tuyl et al., 1988). This method comprises the deposition of pollen on the stylar surface after removing part of the style with stigma and allows pollen to circumvent stylar barriers which normally inhibit pollen tube growth. Once fertilization has occurred, post-fertilization barriers may restrain hybrid embryo growth. In several instances these barriers have been overcome using embryo rescue methods (Asano, 1980). In compatible and interspecific combinations embryos can be excised from immature seeds, starting from about 40 days after pollination, and subsequently cultured and germinated in vitro. Apart from these methods, several additional techniques have been developed for overcoming fertilization barriers in other plant species. Pre-fertilization barriers are bypassed using in vitro pollination and fertilization methods (Zenkteler, 1990), while post-fertilization barriers may be overcome by the culture of ovaries immediately after pollination, or by ovule culture (Guha and Johri, 1966). Aim of our work is to establish an integral system for in vitro pollination, fertilization and embryo rescue and to use this procedure for overcoming both pre- and post- fertilization barriers in wide interspecific *Lilium* crosses.

Application of in vitro methods for genetic manipulation

Broad application of genetic manipulation techniques in monocot plant species is still hindered by the lack of transformation and regeneration procedures. *Agrobacterium* mediated gene transfer has only been successful for a small number of monocots, probably due to the limited host-range of this bacterium. A new method for direct DNA transfer into plant cells has been developed by Klein et al. (1987). The procedure is based on the introduction of genetic material by bombardment of cells or tissues with high velocity microprojectiles using a particle delivery system. Stably transformed maize plants have been obtained by the bombardment of cultured cells (Spencer et al., 1990).

The direct introduction of DNA into pollen would provide an alternative way for transforming plant species. The use of pollen as a delivery vector for foreign DNA to the egg cell was firstly described by Hess (1980). Hess and others attempted to cocultivate *Agrobacterium* together with pollen but were not able to unequivocally prove the transfer of DNA from the bacterium into the male gametophytes. Twell et al. (1989a) used a particle gun for shooting DNA in pollen and confirmed the introduction of foreign DNA in the male gametophytes by establishing -glucuronidase (GUS) marker gene expression in pollen of tobacco. For stable transformation of plants, the transformed pollen has to accomplish fertilization and to transfer the DNA into the egg cell. Goal of our work is to introduce foreign DNA into lily pollen and to use in vitro pollination and fertilization systems for the manipulation of transgenic pollen tubes towards the egg cell.

Material and methods

Plant material

Lilium species, hybrids and cultivars originated from the CPRO plant collection. Species widely differing in flowering period were induced to flower simultaneously by storing bulbs at low temperatures before planting.

In vitro methods

Three types of pollination were used: normal pollination (N), cut- style pollination (CSM) in which pollen was deposited on the cut stylar surface, and grafted style method (GSM) in which pollen was firstly germinated in a compatible style which was cut after 1 day and attached to the ovary of an incongruent mother plant.

N, CSM and GSM pollination were combined with ovary, ovule and embryo culture techniques. The composition of used media has extensively been described by Van Tuyl et al. (1991), and is summarized in Table 1. In short, ovaries or ovary sections were incubated at 250C in the dark, subsequently, ovules were extracted from the ovaries and placed on ovule culture medium. After ovule germination, emerged plantlets were transferred to the embryo rescue medium and grown to little bulbs. After cold treatment, bulbs were planted in soil.

Particle bombardment

Lily (i.e. *Lilium longiflorum*) pollen was collected from mature dehiscent flowers. After rehydration, approximately 70,000 pollen grains were placed on agar plates which were overlayed with a filter paper and a nylon membrane. Bombardment was performed with a DuPont PDS1000 apparatus with tungsten particles coated with DNA. The exact bombardment conditions and DNA precipitation protocols have been described by Van Der Leede-Plegt et al. (in press). Pollen from *Nicotiana glutinosa* was used as a control.

Plasmids

Three plasmids were used in the present study: (I) the pCAL1Gc plasmid provided by Dr. V. Walbot (Stanford University, USA) consisting of the GUS gene fused to a CaMV 35S promoter combined with an ADH1 intron and referred to in the text as 35S, (II) the pCPO1.2'GUS plasmid (referred to as TR2') in which the GUS gene was fused to the 2' dual mannopine synthase gene promoter (Koncz et al., 1990), and, (III) the pLAT52-7 plasmid (referred to as LAT52) provided by Dr. S. McCormick (Plant Gene Expression Centre, USA) containing the flower-specific LAT52 promoter isolated from tomato (Twell et al., 1989b) fused to the GUS reporter gene.

GUS assay

GUS activity was histochemically determined with 2 mM X-gluc (5-bromo-4-chloro-3-indolyl -D-glucuronide) in 150 mM phosphate buffer, pH 7.5. GUS staining was assessed 16 hours after pollen bombardment by applying X-gluc to the nylon membrane carrying the pollen. The number of stained pollen grains was determined after an incubation period of 50 hours.

Table 1. Components supplemented to the MS medium for the different lily culture methods.

Culture	Sucrose%	NAA (mg/l)	pH	DAP
Ovary(-slice)*	9	1	6.0	7-40
Ovule	5	0.1	5.5	42-90
Embryo	2-4	0.001	5.0	40-90
Ovary**	7-8	-	5.8	-2

* Ovaries were transversely cut in 6-8 sections (3-4 mm).
** In vitro culture of whole ovaries.

Results and discussion

Application of in vitro pollination and fertilization for breeding

Methods for in vitro pollination, fertilization and embryo rescue have been optimized for lily. Currently, normal, CSM, and GSM pollination methods can be integrally executed under standard in vitro conditions. With the developed methods, time intervals between pollination and in vitro culture have been reduced from 40 days after pollination (DAP) with embryo rescue to 7 DAP with ovary slice culture (Fig. 1) (Van Tuyl et al., 1991). This makes it possible to overcome pre- and post-fertilization barriers normally impeding the interspecific combinations.

Exploiting CSM and GSM combined with ovary slice method, the in vitro wide interspecific crossing program between different lily species did result in an increase of the number of successfully obtained hybrids. Recently, the newly gained interspecific individuals were tested for various breeding aspects. Apart from new flower colors and forms, a number of the hybrids displayed an ameliorated growth vigour. In general, use of interspecific hybrids in breeding programs may be limited because of a low pollen fertility. Pollen viability of the lily hybrids was tested with fluorescein diacetate (Table 2). While pollen viability of lily species normally ranged from 60-100%, the male vitality of hybrid plants was clearly diminished. Most individuals were completely sterile, although about 7.5% of the hybrids produced pollen which was viable for more than 25%. The viable pollen was analyzed microscopically and with a flow cytometer according to Van Tuyl et al. (1989) and was found to contain the double amount of DNA. Apparently, in most of the lily hybrids meiotic abnormalities induced sterility, while in some individuals meiotic polyploidization resulted in the formation of fertile 2n-pollen. Currently, selected 2n-pollen producing hybrid individuals are crossed with tetraploid lily genotypes to introduce traits from incongruent species into the commercial lily assortment.

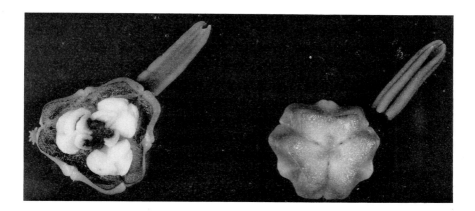

Fig. 1. Ovary slice culture of an Asiatic hybrid, 4 weeks after pollination, on MS medium with 9% sucrose and 1 mg/l NAA, in continuous presence of anthers.

Application of in vitro methods for genetic manipulation

Differential gene expression was found after bombarding 35S, LAT52 and TR2' constructs in *Nicotiana* and lily pollen (Table 3). No blue pollen was observed for both pollen species using the 35S plasmid, reflecting that the regulatory factors necessary for functioning of this CaMV promoter were not expressed in the male gametophytes. Bombardment with LAT52 induced high numbers of *Nicotiana* pollen (2.9%) expressing the marker gene but did not result in stained lily pollen. Apparently, factors activating the LAT52 promoter in pollen of *Nicotiana* were absent in lily. Introduction of TR2' resulted in GUS expression in both pollen species. Conclusively, the TR2' promoter is not only active in vegetative tissues but also in sporogenous cells of both dicot and monocot plants.

The present bombardment results may indicate the feasibility of introducing foreign DNA into lily pollen grains (Fig. 2). However, using the TR2' construct, only about 0.017% of the viable pollen expressed the marker gene. This percentage may for a great part derive from plasmid DNA transiently expressed in the vegetative cell of male gametophytes. For a stable transformation of lily plants, however, the gene construct must be introduced into the nucleus of the generative cell and has to be integrated into the genome of the zygote after fertilization (Heberle-Bors et al., 1990). Clearly, the direct pollination with pollen after bombardment with the TR2' construct will only render a small likelihood for stable transformation. There are at least two alternatives to overcome this problem: (I) the introduction of genes in immature pollen using an in vitro pollen maturation system (Heberle-Bors et al. (1990), and, (II) the introduction of genes in mature pollen but to select for transgenic pollen before pollination. In the last strategy, pollen are to be bombarded with a construct in which a pollen promoter is fused to a selectable marker gene. Subsequently, pollen are selected by supplementing a challenging substance to the pollen germination medium. Transgenic pollen may be separated from non-transgenic individuals based on germinability or pollen tube length (Bino et al., 1987). Using in vitro pollination, limited numbers of selected pollen may effectively accomplish fertilization. Developed embryos may be rescued and be screened in vitro for marker gene expression. In this approach, the combination of in vitro pollination and fertilization with pollen bombardment may give alternative possibilities for the genetic manipulation of lily.

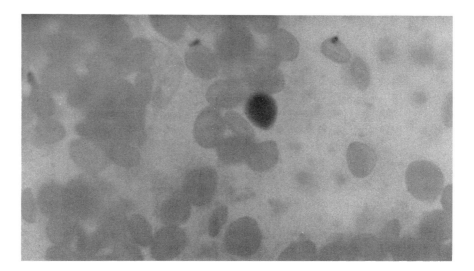

Fig. 2. TR2'-driven GUS expression in *L. longiflorum* pollen.

Table 2. Pollen fertility of lily hybrids obtained from the interspecific in vitro crossing program (n = number of individual hybrid plants).

Cross	Pollen fertility (%)			
	0	0-25	>25	n
L. henryi x L. candidum	2	0	0	2
L. longiflorum x L. candidum	24	2	0	26
L. longiflorum x L. dauricum	22	3	3	28
L. longiflorum x L. henryi	3	1	1	5
L. longiflorum x L. concolor	4	0	1	5

Table 3. Transient GUS expression in Lilium and Nicotiana pollen driven by the various promoters after particle bombardment. GUS activity is shown as the number of blue pollen per 100,000 viable grains and is the mean of three experiments.

Plant	Promoter		
	LAT52	35S	TR2'
N. glutinosa	2904	0	13
L. longiflorum	0	0	17

References

Asano Y (1980) Studies on crosses between distantly related species of lilies. J Jpn Soc Hort Sci 49: 114-118

Ascher PD, Peloquin SJ (1966) Effects of floral aging on the growth of compatible and incompatible pollen tubes in *Lilium longiflorum*. Am J Bot 53: 99-102

Bino RJ, Hille J, Franken J (1987) Kanamycin resistance during in vitro development of pollen from transgenic tomato plants. Plant Cell Reports 6: 333-336

Guha S, Johri BM (1966) In vitro development of ovary and ovule of *Allium cepa* L. Phytomorphology 16: 353-364

Heberle-Bors E, Benito Moreno RM, Alwen A, Stger E, Vincente O (1990) Germ line transformation. In: Progress in plant cellular and molecular biology (Nijkamp HJJ ed) pp. 244-251, Kluwer London

Hess D (1980) Investigations on the intra- and interspecific transfer of anthocyanin genes using pollen as vectors. Z Pflanzenphysiol 98: 321-327

Klein TM, Wolf ED, Wu R, Sanford JC (1987) High-velocity microprojectiles for delivering nucleic acids into living cells. Nature 327: 70-73

Koncz C, Mayerhofer R, Koncz-Kalman Z, Nawrath C, Reiss B, Redei GP, Schell J (1990) Isolation of a gene encoding a novel chloroplast protein by T-DNA tagging in *Arabidopsis thaliana*. EMBO J 9:1337-1346

Spencer TM, Gordon-Kamm WJ, Daines RJ, Start WG, Lemaux PG (1990) Bialaphos selection of stable transformants from maize cell culture. Theor Appl Genet 79: 625-631

Twell D, Klein TH, Fromm ME, McCormick S (1989a) Transient expression of chimeric genes delivered into pollen by microprojectile bombardment. Plant Physiol 91: 1270-1274

Twell D, Wing R, Yuamaguchi J, McCormick S (1989b) Isolation and expression of an anther specific gene from tomato. Mol Gen Genet 217: 240-245

Van Der Leede-Plegt LM, Van De Ven BCE, Bino RJ, Van Der Salm TPM, Van Tunen AJ (in press) Introduction and differential use of various promoters in pollen grains of *Nicotiana glutinosa* and *Lilium longiflorum*. Plant Cell Reports

Van Tuyl JM, Marcucci MC, Visser T (1982) Pollen and pollination experiments. Euphytica 31: 613-619

Van Tuyl JM, Straathof TP, Bino RJ, Kwakkenbos AAM (1988) Effect of three pollination methods on embryo development and seedset in intra- and interspecific crosses between seven *Lilium* species. Sex Plant Reprod 1: 119-123

Van Tuyl JM, De Vries JN, Bino RJ, Kwakkenbos AAM (1989) Identification of 2n-pollen producing interspecific hybrids of *Lilium* using flow cytometry. Cytologia 54: 737-745

Van Tuyl JM, Van Din MP, Van Creij MGM, Van Kleinwee TCM, Franken J, Bino RJ (1991) Application of in vitro pollination, ovary culture, ovule culture and embryo rescue for overcoming incongruity barriers in interspecific *Lilium* crosses. Plant Science 74: 115-126

Zenkteler M (1990) In vitro fertilization and wide hybridization in higher plants. Critical Reviews in Plant Sciences 9: 267-279

POLLEN TUBE ENTRANCE IN THE EMBRYO SAC AND FERTILIZATION

J.L. van Went
Department of Plant Cytology and Morphology
Wageningen Agricultural University
Arboretumlaan 4
6703 BD Wageningen
The Netherlands

In angiosperm plants the interaction between male and female reproductive organs results in double fertilization: the fusion of one sperm cell with the egg cell, and the fusion of the second sperm cell with the central cell. Most of our knowledge of the final stages of the sexual reproduction process, the entrance of the pollen tube in the embryo sac and the actual fertilization events, is based on microscopical observations of fixed and sectioned materials (Jensen 1972; Van Went and Willemse 1984). The embryo sac is positioned deeply inside the ovule, surrounded by a large quantity of untransparant sporophytic tissue, which prevents its direct observation, and the manipulation and experimental analysis of the events that take place. Since recent years it has become possible to separate the embryo sac and its composing cells from the surrounding sporophytic cells by the use of enzymatic maceration techniques (Hu et al. 1985; Zhou and Yang 1986; Mol 1986; Huang and Russell 1989; Wagner et al. 1989; Van Went and Kwee 1990; Kranz et al. 1991). These techniques allow the isolation of embryo sacs and composing cells in large quantities and in living and intact condition. Presently, in a number of laboratories molecular and cell biological research is in progress focused on the developmental and functional aspects of the female gametophyte and gametes. Also sperm cells can be isolated in living condition and large quantities (Theunis et al. 1991). Isolated male and female gametes are now used to unravel the basic principles of gamete fusion (Kranz et al. 1991).

From microscopical studies we have learned that the compatible pollen tube grows along the surface of the placenta towards the ovules. In some species special structures, like an obturator or papillate placental cells, provide a well-defined pathway which guides the pollen tube to the apex of an ovule. To reach the embryo sac the pollen tube normally enters the micropyle. It is generally assumed that this stage of pollen tube growth is directed chemotropically by substances produced by the ovule and secreted through the micropyle. The synergids of the embryo sac are considered to produce such chemotropic substances, and to excrete them into the filiform apparatus. From the filiform apparatus the substances are supposed to leach out into the micropyle or the adjacent nucellus cells, creating a gradient that influences the growth direction of the pollen tube. In species with tenuinucellate ovules, the mature embryo sac is not surrounded by nucellus tissue, and the pollen tube directly meets the embryo sac apex after passage of the micropyle. In species with crassinucellate ovules the mature embryo sac is surrounded by nucellus tissue. The pollen tube has to enter this nucellus tissue, after it has passed the micropyle in order to reach the embryo sac. In these cases, it has been found that more then one pollen tube can reach the nucellus, but only one pollen tube actually penetrates. Apparently, a recognition and blocking system is operating here, leading to the rejection of supernumerous pollen tubes.

Pollen tube entrance in the embryo sac

Since the 1960's, a large number of articles has been published on the ultrastructural aspects of the entrance and discharge of the pollen tube in the embryo sac (Van Went and Willemse 1984). Invariably, it has been observed that the pollen tube enters the embryo sac at the micropylar side, where it penetrates through the filiform apparatus into one of the synergids. It never enters the egg cell or the central cell. Even in species that do not have synergids, like Plumbago zeylanica, the pollen tube approaches the embryo sac at the micropylar side and it delivers the sperm cells without distortion of the egg cell and the central cell (Russell 1982). In the latter species the portion of the wall at the micropylar apex of the embryo sac and the adjacent portion of the wall in between the egg cell and the central cell highly resembles the filiform apparatus formed by the synergids in other species. And it is into this specialized cell wall part that the pollen tube penetrates, as it does into the filiform apparatus of the synergids. Evidently, the filiform apparatus or a comparable structure plays a key role in the interaction between the male and the female gametophyte. It is not known in which way the pollen tube penetrates the filiform apparatus, nor what causes it to penetrate. Although there appears to be considerable variation in morphology and ultrastructure of the filiform apparati among species, in general they are considered to be largely composed of pectines in which a loosely organized cellulosic microfibrillar network is embedded. In this respect the filiform apparatus clearly differs from the neighbouring cell walls. It might be that the specific composition and physical properties of the filiform apparatus allow the pollen tube to penetrate by mechanical forces. Another possibility is that the pollen tube excretes enzymes which tranform the filiform apparatus into a suitable and attractive pathway for the pollen tube. It is known that growing pollen tubes in the style produce and secrete cell wall degrading enzymes, like pectinases, which are involved in the breakdown of the intercellular substance of the transmitting tissue.

It has been postulated that the filiform apparatus acts as a source of substances that chemotropically forces the pollen tube to grow towards the embryo sac. Thus far no direct evidence has been found to support this hypothesis. Experiments to establish a directional effect of isolated ovules on pollen tube growth have been unsuccessful. The main arguments for the chemotropy hypothesis are based on functional interpretations of the cytoplasmic ultrastructure and polar organization of the synergids. Most of the synergid cytoplasm is located in the micropylar half of the cell near the filiform apparatus. The cytoplasm has large numbers of mitochondria and ribosomes, and an extensive rough endoplasmic reticulum in stacked configuration. These characteristics are interpreted as signs of high metabolic activity and synthesis of products to be secreted. The location of this activity, in combination with the transfer wall-like character of the filiform apparatus points to secretion of the produced substances or their derivatives into the filiform apparatus. Likely, future molecular and cell biological research using isolated synergids, will be more successful in elucidating this topic.

After the pollen tube has passed the filiform apparatus it enters one of the synergids. In many species one of the synergids degenerates before the arrival of the pollen tube. Invariably, it is into this degenerated synergid that the pollen tube enters. In other species, however, both synergides degenerate before the pollen tube arrives, or both synergides remain intact until the moment of penetration.

It is not clear what factors determine which of the synergids is to be penetrated by the pollen tube. Soon after the pollen tube has reached the synergid cytoplasm, its growth ceases and the tube opens. In some species it appears that the tube simply bursts at its tip. Bursting could result from a sudden change in osmotic conditions, or could be induced by oxygen depletion, or could be triggered by specific chemicals or enzymes in the synergid. In other species, however, a distinct subterminal pore is formed in the pollen tube, and cessation of tube growth and opening of the tube may be separated processes.

Subsequently a considerable portion of the tube content, including the vegetative nucleus and the two sperm cells, is injected into the penetrated synergid. The injection is accompanied or followed by the collapse of the synergid vacuole, the swelling of the cell, and a number of degenerative changes of the cytoplasm. Frequently, a degeneration or rupture of the synergid plasma membrane in the chalazal region of the cell is observed, and degenerated cytoplasm of the penetrated synergid invades in between the adjacent cells. In this part of the embryo sac the neighbouring cells (synergids, egg cell and central cell) are separated from each other by their plasma membranes only, and solid cell walls are absent. But apparently, the plasma membranes of the egg cell and central cell are protected in some way from being affected by the contact with the degenerating synergid cytoplasm. Degenerative changes also occur in the injected pollen tube material, except in the sperm cells. Inside the pollen tube, a sperm cell is separated from the vegetative cell only by the two plasma membranes and a very thin intercellular layer. Inside the synergid, the vegetative plasma membrane which surrounds the sperm cell rapidly degenerates, but apparently the sperm cell itself is protected in some way, like the egg cell and the central cell. The sperm cells, however, remain not unaltered. Inside the pollen tube the sperm cells have a clear spindle shape, and they contain numerous bundles of cortical microtubules. Inside the penetrated synergid, the shape of the sperm cells changes to spherical, and the cortical microtubules disappear (Mogensen 1982). In these respects they strongly resemble sperm cells after isolation from pollen grains or tubes (Theunis 1990). In species like *Plumbago zeylanica* which lack synergids, the pollen tube injects its content into the filiform apparatus, from which it invades in between the egg cell and the central cell. The injected material shows similar degenerative changes as are observed in penetrated synergids.

Gamete fusion

As result of the events described before, at the time of actual fertilization the male and female gametes are transformed in complete or partial naked protoplasts, and the plasma membranes of the sperm cells become directly appressed to the plasma membranes of the egg cell and the central cell. In what way the gametes approach each other is not known. It might be that the mechanical forces of the injection process cause the sperm cells to move to their final positions for the subsequent fusion. Likely, in the pollen tube the sperm cells are subjected to transport based on an actin-myosin mechanism located in the vegetative cell (Heslop-Harrison and Heslop-Harrison 1989). But it is unlikely that such a mechanism is operating in the penetrated and degenerating synergid. Furthermore, it is unlikely that angiosperm sperm cells can move freely themselves, since independent and active movement of isolated sperm cells never has been observed. The mechanism of sperm cell movement becomes even more intriguing since it has been established that, at least in some species with dimorphic sperm cells, mating of the gametes is not at random. In these species preferential fertilization occurs, in which one sperm cell is predetermined to fuse with the egg cell, while the second sperm cell always fuses with the central cell (Russell 1985). This preferential fertilization likely requires an adequate recognition system, by which the female gametes can recognize their respective fusion partners. Another possibility could be a very precise deposition mechanism, by which predetermined fusion partners could be brought together.

Gametic fusion occurs very rapidly. In *Plumbago zeylanica* it has been established that the presence of unfused sperm cells in the female gametophyte is lasting less than 5 minutes over a period of 8.5 hr of pollen tube growth (Russell 1983). Fusion of the male and female gametes starts with local contact and fusion of their respective membranes. In *Plumbago zeylanica*, gametic fusion is initiated by a single fusion event between the membranes of the male and female gametes, followed by secondary fusions at different locations along their surfaces. Evidently, gametic fusion in angiosperms is a membrane-based event, for which the characteristics, exposure and contact of the gametic membranes are critical. As is pointed out earlier, exposure and contact of the gametic membranes is facilitated by the degeneration of the synergid and pollen tube cytoplasm and plasma membranes, and the bursting of the penetrated synergid. The degenerative changes in the penetrated synergid are not just a result of the pollen tube entrance and the interaction between the pollen tube and synergid content, but they appear to be vital prerequisites for successful fertilization. In addition, the gametic membranes must have specific characteristics to facilitate the membrane recognition and the membrane fusion. Such specific characteristics have been observed in sperm cell plasma membranes by freeze-fracture analysis (Van Aelst et al. 1990). With this technique it is possible to establish the number, position and arrangement of intramembrane protein particles (IMPs). Generally, in somatic plant cell plasma membranes the cytoplasmic half of the membrane (PF face) contains more IMPs than the peripheral half (EF face). In sperm cell plasma membranes the EF face appears to contain three times more IMPs than the PF face. This high number of IMPs in the EF face of the sperm cell plasma membrane could well be related to the specific recognition and fusion capacities of the gametic plasma membranes.

The membrane-based fusion concept implicates that not only the male gamete nucleus is transmitted, but that there is also co-transmission of the sperm cell cytoplasmi. Co-transmission of sperm cell cytoplasm has well been established by genetical studies using marker genes located in male cytoplasmic organelles (Kirk and Tilney-Bassett 1978).

The next step in the process of double fertilization is the fusion of the sperm cell nucleus with the egg cell nucleus, and the fusion of the second sperm cell nucleus with the (fused) polar nuclei. The nuclear fusions start with local contacts and fusions of subsequently the outer nuclear membranes and the inner nuclear membranes (Jensen 1964; Mogensen 1982). In this way nucleoplasmic bridges are formed between the fusing nuclei, which enlarge and coalesce. Finally the contents of the male and female nuclei intermingle, which marks the completion of fertilization.

Recommended literature

Heslop-Harrison J, Heslop-Harrison Y (1989) Myosin associated with the surfaces of organelles, vegetative nuclei and generative cells in angiosperm pollen grains and tubes. J Cell Sci 94:319-325

Hu S-Y, Li L-G, Zhou C (1985) Isolation of viable embryo sacs and their protoplasts of *Nicotiana tabacum*. Acta Bot Sin 27:337-344

Huang B-Q, Russell SD (1989) Isolation of fixed and viable eggs, central cells and embryo sacs from ovules of *Plumbago zeylanica*. Plant Physiol 90:9-12

Jensen WA (1964) Observations on the fusion of nuclei in plants. J Cell Biol 23:669-672

Jensen WA (1972) The embryo sac and fertilization in angiosperms. Harold L. Lyon Arbor Lect 3:1-32

Kirk JTO, Tilney-Bassett RAE (1978) The plastids: their chemistry, structure, growth and inheritance. Elesevier/North Holland Biomed Press, Amsterdam

Kranz E, Bautor J, Lrz H (1991) In vitro fertilization of single, isolated gametes of maize mediated by electrofusion. Sex Plant Reprod 4:12-16

Mogensen HL (1982) Double fertilization in barley and the cytological explanation for haploid embryo formation, embryoless caryopses, and ovule abortion. Carlsberg Res Commun 47:313-354

Mol R (1986) Isolation of protoplasts from female gametophytes of *Torenia fournieri*. Plant Cell Rep 3:202-206

Russell SD (1982) Fertilization in *Plumbago zeylanica*: entry and discharge of the pollen tube in the embryo sac. Can J Bot 60:2219-2230

Russell SD (1983) Fertilization in *Plumbago zeylanica*: gametic fusion and fate of the male cytoplasm. Amer J Bot 70:416-434

Russell SD (1985) Preferential fertilization in *Plumbago*: ultra-structural evidence for gamete-level recognition in an angiosperm. Proc Natl Acad Sci, USA 82:6129-6132

Theunis CH (1990) Ultrastructural analysis of *Spinacia oleracea* sperm cells isolated from mature pollen grains. Protoplasma 158:176-181

Theunis CH, Pierson ES, Cresti M (1991) Isolation of male and female gametes in higher plants. Sex Plant Reprod 4:145-154

Van Aelst AC, Theunis CH, Van Went JL (1990) Freeze-fracture studies on isolated sperm cells of *Spinacia oleracea* L. Protoplasma 53:204-207

Van Went JL, Willemse MTM (1984) Fertilization. In: Johri BM (ed) Embryology of angiosperms. Springer, Berlin Heidelberg New York Tokyo, pp 273-317

Van Went JL, Kwee H-S (1990) Enzymatic isolation of living embryo sacs of *Petunia*. Sex Plant Reprod 3:257-262

Wagner VT, Kardolus JP, Van Went JL (1989) Isolation of the lily embryo sac. Sex Plant Reprod 2:219-224

Zhou C, Yang HY (1986) Isolation of embryo sacs by enzymatic maceration and its potential in haploid study. In: Hu SY, Yang HY (eds) Haploids of higher plants in vitro. Springer, Berlin Heidelberg New York, pp 192-203

PREFERENTIAL FERTILIZATION: DATA AND STRATEGY FOR MOLECULAR ANALYSIS

A. CHABOUD
Reconnaissance Cellulaire et Amélioration de Plantes
Université LYON 1
43 bd du 11 novembre 1918
69622 Villeurbanne Cedex
France

Double fertilization was first described by Nawashin (1898) and Guignard (1899) and involves the fusion of the two male gametes with the two female gametes within the embryo sac: one fusion initiates embryo formation and a second fusion triggers formation of the secondary endosperm (reviewed in Knox et al., 1986).

The two fusions, characteristic of double fertilization, occur at adjacent cells close together in time and without competition from other male gametes since in most cases, only one pollen tube discharges into a synergid (reviewed in Knox et al., 1986). Transmission electron microscopy studies evidence that the male gametes are discharged free from the outer vegetative membrane and that fusion occur through interrupted cell walls in the female gametophyte (Jensen, 1974; Mogensen, 1978,1982; Russell, 1983). So, it is reasonable that higher plant gametes might have cell surface determinants that function in fertilization as lower plant gametes or animal gametes.

The process of double fertilization previously described may function according to two alternatives: a "random" or a "directed phenomenon". In the random hypothesis, either one or the other male gamete can equally fuse with one or the other female target cell, suggesting that the two gametes of one pollen grain are identical. In the case of a directed phenomenon, each male gamete of one pollen grain would be "programmed" for preferential fusion with either the egg cell or the central cell, suggesting that gametic recognition occurs.

The ultimate aims of this paper is, after a description of the two examples showing preferential fertilization in Angiosperms, to point out new developments within the last five years (access to male and female gametes isolated from their gametophytes), which offer new strategy to test the hypothesis of preferential fertilization and to analyze gametic recognition at the molecular level.

Preferential fertilization: fact or hypothesis ?

In the litterature, the hypothesis of preferentialis supported by only two examples where male gamete dimorphism.was associated to demonstration of preferential fusion.

Male gamete nuclear dimorphism was genetically established in certain lines of maize by Roman (1948), in which B-chromosomes were frequently (frequency = 50 to 98 %) transmitted unequally into the sperm descendents of the generative cell. Some male gametes of maize would contain two B-chromosomes (or more) and its "sister" cell would contain less than its expected complement. This form, since termed "nuclear heterospermy" (Russell, 1985), was also linked to functional differences since the male gamete with the excess complement of B-chromosomes was genetically more likely (frequency = 66 %) to fuse with the egg cell. Roman (1948) attributed "directed fertilization", as he termed it, to order of sperm cell arrival in the female gametophyte.

The second example concerns a cytological evidence of preferential fertilization in *Plumbago zeylanica* , in which male gamete are dimorphic in terms of plastids and mitochondria content (Russell, 1985). Consequently, the fate of maternal and paternal organelles can be traced by organellar differences and therefore, not only is the fate of the dimorphic male gametes known, but the participation of their cytoplasm is well documented (Russell, 1983). In this case, preferential fertilization results in the fusion of the plastid-rich gamete into the embryo in over 94% of the cases, strongly supporting the presence of recognition factors that may recognize and discriminate between the male gametes of this plant. However, according to currently available data (Corriveau and Coleman, 1988), this strong pattern of plastid dimorphism seems restricted to the immediate subfamily to which *Plumbago* belongs.

Differences in the nuclear genetics of male gametes are extremely rare, since they must occur during generative cell mitosis when the gametes form. More common is the situation of "cytoplasmic heterospermy". Nevertheless, the majority of the flowering plants studied to date, express a weak form of cytoplasmic dimorphism (reviewed in Knox *et al.*, 1988) in which the two male gametes may differ in size, shape and mitochondrial content, but not plastid content. Whether these express preferentiality in fertilization remains to be demonstrated. Moreover, evidence in barley suggests that fertilization may exclude the male cytoplasm in entirety (Mogensen, 1988). In these plants and others in which the male gametes do not appear dimorphic, demonstration of any preferentiality of fertilization between the two male gametes will be essentially impossible.

Now, it appears that the only alternative to answer the question of preferential fusion and gametic recognition during double fertilization is to get direct access to gametic interactions *via* isolation of male and female gametes.

What is the best model for gametic interactions analysis ?

In order to study gametic interactions, one needs to work with the largest amounts of intact, viable and functional male and female gametes isolated from their gametophytes. In this view, several attempts have been carried out for different species in the last decade

Female gametes isolation requires essentially an enzymatic digestion followed by some mechanical disruption to ensure proper liberation of embryo sac (Table 1). After this enzymatic treatment, the method commonly used to test viability is based on the evaluation of both membrane integrity and enzyme activity, the fluorochromatic reaction test (FCR, Heslop-Harrison and Heslop-Harrison, 1970). Then, it is possible to microdissect the viable isolated embryo sac to obtain the female target cells of the male gametes, egg cell and central cell. The yield is mostly rather low (*circa* 5 to 10%) but nevertheless satisfactory for maize (for example, 100 to 200 living isolated embryo sacs per day and per experimentor).

SPECIES	ISOLATION METHOD	VIABILITY	REFERENCES
Dicotyledons			
Nicotiana tabacum	enzymatic digestion, squashing	+ (FCR)	Hu *et al.*, 1985
Antirrhinum majus	enzymatic digestion	+ (FCR)	Zhou and Yang, 1985
Torenia fournieri	enzymatic digestion	+ (FCR)	Mol, 1986
Helianthus annuus	enzymatic digestion	+ (FCR)	Zhou, 1987
Plumbago zeylanica	enzymatic digestion, dissection	+ (FCR)	Huang *et al.*, 1989
Monocotyledons			
Lilium longiflorum	enzymatic digestion, suction	+ (FCR)	Wagner *et al.*, 1989b
Zea mays	enzymatic digestion, dissection	+ (FCR, TEM)	Wagner *et al.*, 1988 Wagner *et al.*, 1989c

FCR: fluorochromatic reaction test (according to Heslop-Harrison & Heslop-Harrison, 1970)
TEM: transmission electron microscopy

TABLE 1: ISOLATION OF VIABLE EMBRYO SACS IN DIFFERENT SPECIES

For male gametes, as they are embedded in the vegetative cells of pollen grains or pollen tubes, their release can be obtained by the aid of an osmotic shock, grinding the pollen grain, or by wall degrading enzymes (for an analysis of different protocols, see Roeckel *et al.*, 1990a). The methods used must be adapted to the species studied, to allow pollen grains to break without damaging male gametes. Another step of the isolation procedure is to separate released male gametes from organelles and cytoplasmic debris. Larger contaminants can be removed by filtration. Smaller contaminants can be separated from male gametes by gradient centrifugation (for an analysis of different protocols, see Roeckel *et al.*, 1990a). Nevertheless, for further cellular recognition studies, the main key point is to isolate male gametes in large numbers and of good quality. Comparison of the results obtained with several species (Table 2) clearly indicate that maize is the best suited species in this view.

SPECIES	YIELD [a]	VIABILITY[b]	REFERENCES
Dicotyledons			
Plumbago zeylanica	$1.7\ 10^5$ cells, 60 to 75%	95%, (Ev B-)	Russell, 1986
Rhododendron spp.	90 to 270 pairs per style [c]	nd	Shivana *et al.*, 1988
Brassica oleracea	2%	nd	Roeckel *et al.*, 1988
Beta vulgaris	$7\ 10^4$ cells	30%, (FCR+)	Nielsen and Olesen,1988
Gerbera jamesonii	$6\ 10^4$ cells	nd	Southworth and Knox, 1989
Spinacia oleracea	10^5 cells, 5 to 10%	90%, (FCR+)	Theunis and Van Went,1989
Brassica napus	$2.7\ 10^4$ cells	90%, (TEM)	Taylor *et al.*, 1991
Monocotyledons			
Zea mays	$3\ 10^6$ cells, 20 to 30 %	80-90%, (FCR+)	Dupuis *et al.*, 1987 Roeckel *et al.*, 1988
Zea mays	$1.5\ 10^6$ cells	50%, (Ev B-)	Cass and Fabi, 1988
Gladiolus gandavensis	65 to 84 pairs per style [c]	nd	Shivanna *et al.*, 1988
Lolium perenne	2%	nd	Van der Maas and Zaal, 1990

(a): Yield was reconsidered from quantitative data of papers, as total number of isolated cells obtained at the end of a single isolation procedure and/or percentage of isolated cells recovered from the initial number of male gametes within pollen grains.
(b): Viability percentage of isolated cells at the end of isolation procedure. FCR+: positive fluorochromatic reaction using fluorescein diacetate; EvB-: Evans blue excluded; TEM: transmission electron microscopy, nd: not determined.
(c): In this study, male gametes have been isolated from pollen tubes growing in style segments by the *in vivo/in vitro* method, and yield has been estimated by the number of isolated pairs of male gametes per style.

TABLE 2: ISOLATION OF MALE GAMETES IN DIFFERENT SPECIES. COMPARISON OF YIELD OBTAINED AND VIABILITY OF ISOLATED CELLS.

In conclusion, this rapid survey indicate that maize emerges as the species of choice to isolate the largest quantity of viable female and male gametes. Moreover, maize is the flowering plant among the best suited for biological research, because of its economic importance as well as the considerable amount of genetic and cytogenetic informations currently available (Wessler and Hake, 1990; Walbot,

1991). In addition, the recent development of molecular biology has provided a lot of valuable RFLPs informations as well as isolation of pollen-specific genes (reviewed in Mascarenhas, 1990). The second key point for this choice is related to its reproductive biology itself. Maize bears its male and female flowers on separate structures that makes easy the collection of both pollen and pistil. Moreover, several grams of pollen grains can be produced by a single tassel.and its large ear with 500 or more individual spikelets is a good starting material to obtain large number of isolated embryo sacs.

Consequently, our group has chosen maize as a model, and succeeds in isolation of intact, viable and functional maize gametes (Dupuis *et al.*, 1987; Wagner *et al.*, 1989c). Then, we have used numerous methods for assessing the quality and the functional state of isolated male gametes. Transmission electron microscopy was used (Dupuis *et al.*, 1987) to allow direct visualisation of both plasma membrane configuration and intactness, but also the observable condition of cellular organelles (Wagner *et al.*, 1989a). The functional state of the plasma membrane was demonstrated by measurement of transmembrane currents recorded by preliminary patch-clamp experiments in cell attached configuration (Roeckel *et al.*, in preparation). Further studies have shown the presence of ATP within isolated male gametes from *Zea mays* indicating a metabolic potential of these cells (Roeckel *et al.*, 1990b). Moreover, ^{35}S-methionine labelling experiments clearly indicate that these isolated cells are also able to synthesize new proteins (Roeckel *et al.*, in preparation). However, whether isolated male gametes remain able to fertilize the egg or central cell will be the ultimate biotechnological assay.

Strategy for gametic interactions analysis

Our success in isolation of intact, viable and functional maize gametes offers the possibility of a direct access to the interacting gamete membranes supposed to harbor the specific determinants of interest in double fertilization. Thus, we are currently designing an experimental strategy based on the construction of a monoclonal antibody library directed against the cell surface of isolated viable male gametes, and the development of a model of *in vitro* intergametic fusion

Monoclonal antibodies were first raised to isolated sperm of *Brassica campestris* pollen (Hough *et al.*, 1986) but more detailed investigations on the potential use of monoclonal antibodies for male gametes characterisation were reported by Pennell *et al.*, (1987). In this study, male gametes of *Plumbago zeylanica* appeared to be strongly immunogenic as a good proportion of reactive lines were sperm-specific. However, the strongly reactive lines were directed against nuclear or cytoplasmic components of the gamete, and many more reacted with distinctly particulate or soluble components of pollen cytoplasm. This study clearly indicates that more purified male gamete preparations free of pollen cytoplasm.contaminants will be necessary to ellaborate more discriminating screening procedures to select gamete-specific cell-surface determinants. Similarly, comparison by electrophoretic methods of protein patterns of homogenates of isolated male gametes and respective pollen preparations (Knox *et al.*, 1988; Geltz and Russell, 1988) suggests that a high degree of overlap exists between gamete-rich and pollen fractions. Our own first trials of mice immunizations pointed out the same limiting factor, so that the large scale purification of isolated male gametes is the objective currently carried on in the lab. However, as in maize, gametes volume represents only 0.02 to 0.2% of the pollen volume, this may require considerable effort to get purified fractions suitable for analysis

Nevertheless, as soon as this problem will be solved, the monoclonal antibody library directed against the cell surface of isolated male gametes will be constructed and used for cell-sorting the population of isolated gametes. This will allow to assess membrane dimorphism of the two male gametes of one pollen grain. The development of an *in vitro* model of intergametic fusion without any artificial fusion inducer currently in progress will be used for experiments of fusion inhibition by monoclonal antibodies. This would allow us to sort specific cell surface determinants involved in gametic recognition in maize.

This approach should lead in the future to the study of the ontogenesis and the role in double fertilization process of the gamete-specific determinants identified.

Acknowledgements:
Most of the work reviewed here was carried out in the group of Pr C. DUMAS with the financial support of EEC (Biotechnology Action Program 0203-F) and INRA that are gratefully acknowledged.

LITTERATURE REFERENCES

CASS, D.D. and FABI, G.C. (1988). Structure and properties of sperm cells isolated from the pollen of *Zea mays* . Can. J. Bot., 66, 819-825

CORRIVEAU, J.L. and COLEMAN, A.W. (1988). A rapid screening method to detect potential biparental inheritance of plastid DNA and results for over 200 selected angiosperms. Amer. J. Bot. 75, 1443-1458

DUPUIS, I., ROECKEL, P., MATTHYS-ROCHON, E., DUMAS, C. (1987). Procedure to isolate viable sperm cells from corn (*Zea mays* L.). pollen grains. Plant Physiol. 85, 876-878

GELTZ, N.R. and RUSSELL, S.D. (1988). Two dimensional electrophoretic studies of the proteins and polypeptides in mature pollen grains and the male germ unit of *Plumbago zeylanica* . Plant Physiol. 88, 764-769

GUIGNARD, L. (1899). Sur les anthérozoides et la double copulation sexuelle chez les végétaux angiospermes. Rev. Gén. Bot. 11, 129-135

HESLOP HARRISON, J. and HESLOP HARRISON, Y. (1970). Evaluation of pollen viability by enzymatically induced fluorescence, intracellular hydrolysis of fluorescein diacetate. Stain Technol. 45, 115-120

HOUGH, T., SING, M.B., SMART, I.J. and KNOX, R.B. (1986). Immunofluorescent screning of monoclonal antibodies to surface antigens of animal and plant cells bound to polycarbonate membranes. J.Immunol. Methods, 92, 103-107

HU, S.Y., LI, L.G. and ZHU, C. (1985). Isolation of viable embryo sacs and their protoplasts of *Nicotiana tabacum* . Acta Bot. Sin., 27, 337-344.

HUANG, B.Q., STROUT, G.W. and RUSSELL, S.D. (1989). Isolation of fixed and viable eggs, central cells, and embryo sacs from ovules of *Plumbago zeylanica*.. Plant Physiol. 90, 9-12.

JENSEN, W.A. (1974). Reproduction in flowering plants. In "Dynamic aspects of plant ultrastructure. (eds Robards, A.W) pp.481-503 McGraw-Hill New York

KNOX, R.B., SOUTHWORTH, D. and SINGH, M.B. (1988). Sperm cells determinants and control of fertilization in plants. In "Eucaryote cell recognition : concepts and model systems". (eds CHAPMAN, G.P., AINSTHWORTH, C.C. and CHATHAM, C.J.) pp. 175-193. Cambridge University Press

KNOX, R.B., WILLIAMS, E.G. and DUMAS, C. (1986). Pollen, pistil and reproductive function in crop plants. Plant Breed. Rev. 4, 8-79

MASCARENHAS, J.P. (1990). Anther and pollen expressed genes. Ann. Rev. Plant Physiol. Plant Mol. Biol. 41, 317-338.

MOGENSEN, H.L. (1978). Pollen tube-synergid interactions in *Proboscidea louisianica* (Martineaceae). Amer. J. Bot. 65,953-964

MOGENSEN, H.L. (1982). Double fertilization in barley and the cytological explanation for haploid embryo formation, embryoless caryopses, and ovule abortion. Carlsberg Res. Com. 47, 313-354

MOGENSEN, H.L. (1988). Exclusion of male mitochondria and plastids during syngamy as a basis for maternal inheritance. PNAS (USA) 85, 2594-2597

MOL, R. (1986). Isolation of protoplasts from female gametophytes of *Torenia fournieri*. Plant Cell Rep., 3, 202-206.

NAWASHIN, S.G. (1898). Resultate einer Revision der Befruchtungs von gänge bei *Lilium martagon* and *Fritillia tenella*. Izo. Imp. Akad. Nank. 9, 377-382

NIELSEN, J.E. and OLESEN, P. (1988). Isolation of sperm cells from the trinucleate pollen of sugar beet (*Beta vulgaris*). In "Plant sperm cells as tools for biotechnology", (eds WILMS, H.J. and KEIJZER, C.J.), pp. 111-112 Pudoc Wageningen

PENNELL, R.I., GELTZ, N.R., KOREN, E. and RUSSEL, S.D. (1987). Production and partial characterization of hybridoma antibodies elicited to the sperm of *Plumbago zeylanica*.. Bot. Gaz., 148, 401-406.

ROECKEL, P., CHABOUD, A., MATTHYS-ROCHON, E., RUSSELL, S. and DUMAS, C. (1990a). Sperm cell structure, development and organisation. In: Microspores. Evolution and ontogeny (Blackmore S. and Knox R.B., Eds), Academic Press, London. pp: 281-307.

ROECKEL, P., DUPUIS, I., DETCHEPARE, S., MATTHYS-ROCHON, E. and DUMAS, C. (1988). Isolation and viability of sperm cells from corn (*Zea mays*) and kale (*Brassica oleracea*) pollen grains. In "Plant sperm cells as tools for biotechnology" (eds WILMS, H.J. and KEIJZER, C.J.). pp. 105-110 Pudoc Wageningen

ROECKEL, P., MATTHYS-ROCHON, E. and DUMAS, C. (1990b). Pollen and isolated sperm cell quality in *Zea mays*. In "Characterization of male transmission units in higher plants" (ed, BARNABAS, B. and LISZT, K.) pp.41-48. MTA Copy, Budapest.

ROMAN, H. (1948). Directed fertilization in maize. PNAS (USA) 34, 36-42

RUSSELL, S.D. (1983). Fertilization in *Plumbago zeylanica* : gametic fusion and fate of the male cytoplasm. Amer. J. Bot. 70, 416-434

RUSSELL, S.D. (1985). Preferential fertilization in *Plumbago* : ultrastructural evidence for gamete-level recognition in an angiosperm. PNAS (USA) 82, 6129-6132

RUSSELL, S.D. (1986). Isolation of sperm cells from the pollen of *Pumbago zeylanica*. Plant Physiol., 81, 317-319.

SHIVANNA, K.R., XU, H., TAYLOR, P. and KNOX, R.B. (1988). Isolation of sperms form the pollen tubes of flowering plants during fertilization. Plant Physiol., 87, 647-650

SOUTHWORTH, D. and KNOX, R.B. (1989). Flowering plant sperm cells: Isolation from pollen of *Gerbera jamesonii* (Asteraceae). Plant Sci., 60, 273-277.

TAYLOR, P., KENRICK, J., BLOMSTED, C.K. and KNOX, R.B. (1991). Sperm cells of the pollen tubes of *Brassica* : Ultrastructure and isolation. Sex. Plant Reprod. 4, 226-234.

THEUNIS, C.H. and VAN WENT, J.L. (1989). Isolation of sperm cells from mature pollen grain of *Spinacia oleracea* L.. Sex. Plant Reprod., 2, 97-102.

VAN DER MAAS, H.M. and ZAAL, M.A.C.M. (1990). Sperm cell isolation from pollen of perennial ryegrass (*Lolium perenne* L.). In "Characterization of male transmission units in higher plants" (ed, BARNABAS, B. and LISZT, K.) pp.31-36. MTA Copy, Budapest.

WAGNER, V.T., SONG, Y.C., MATTHYS-ROCHON, E. and DUMAS, C. (1988). The isolated embryo sac of *Zea mays* : structural and ultrastructural observations.In "Sexual reproduction in higher plants". (eds CRESTI, M., GORI, P. and PACINI, E.) pp.233-238. Springer Verlag, Berlin Heidelberg.

WAGNER, V.T., DUMAS, C. and MOGENSEN, H.L. (1989a). Morphometric analysis of isolated *Zea mays* sperm. J. Cell Sci., 93, 179-184.

WAGNER, V.T., KARDOLUS, J.P. and VAN WENT, J.L. (1989b).Isolation of the lily embryo sac. Sex. Plant Reprod., 2, 219-224.

WAGNER, V.T., SONG, Y.C., MATTHYS-ROCHON, E. and DUMAS, C. (1989c). Observations on the isolated embryo sac of *Zea mays* . Plant Sci., 59, 127-132.

WALBOT, V. (1991). Maize mutants for the 21st century. Plant Cell, 3, 851-856.

WESSLER, S. and HAKE, S. (1990). Maize harvest. Plant Cell, 2, 495-499.

ZHOU, C. and YANG, H.Y. (1985). Observations on enzymatically isolated, living and fixed embryo sacs in several angiosperm species. Planta, 165, 225-231.

ZHOU, C. (1987). A study of fertilization events in living embryo sacs isolated from sunflower ovules. Plant Sci. 52, 147-151

EFFECTS OF GAMETOPHYTIC SELECTION ON THE GENETIC STRUCTURE OF POPULATIONS

M. Sari Gorla
Department of Genetics and Microbiology
University of Milano
Via Celoria 26, 20133 Milano
ITALY

1. Gametophytic selection: what it is

In the classical view, population genetic models are based on the assumption that the haploid phase is genetically silent; thus they take into account the adaptive characteristic only of the diploid phase. This is likely to be true of animal populations: notwithstanding many examples of non-Mendelian segregation have been described (Zimmering et al 1970), in animals the opportunities for gametic selection are small, since the phenotypes of eggs and sperms seem to be largely determined by the diploid genotypes of the organism that produced them. In lower plants, such as mosses, in which the gametophytes are a major part of the life cycle, selection in this stage is certainly important, perhaps more so than is selection in sporophytes. In higher plants the issue is not so evident, but it is now well established that the conventional assumption is not justified: when the effects of selection in natural or cultivated plant populations are studied, not only sporophytic, but also gametophytic selection has to be taken into account.

Gametophytic selection refers to differential gene transmission during the haploid phase of higher plants, from meiosis to fertilization. This essentially concerns the male gametophyte, rather than the female, due to its peculiar characteristics: large population size, independence of the maternal plant, direct exposure to environmental factors, a large number of individuals competing with each other during development, germination and tube growth, within the same style, to achieve fertilization. Moreover, since pollen is a very simple, haploid structure, also recessive alleles are exposed to the selective effects. These special features of the male gametophyte indicate that selection in this phase can be much more effective than selection on the sporophyte, and suggest that it plays a central role in the evolution of higher plants.

The assumption with regard to non-random fertilization is that in pollen there is expression of genes the alleles of which are able to confer different adaptive values, such as genes controlling traits related to pollen fitness components or resistance to environmental stresses. Consequences of differential gene transmission are the increase of the favourable allele frequency of the genes responsible for the trait and of the linked genes, and the deviation from random mating. Demonstrations of haploid gene expression have been given both for genes used as "genomic markers" (genes coding for isozymes, for instance) and for genes involved in adaptive trait control, such as genes for HSPs (heat shock proteins), likely to be involved in conferring thermotolerance (Frova et al 1989), genes controlling pollen competitive ability (Ottaviano et al 1988), or *in-vitro* tube growth rate (Sari-Gorla et al 1975), and genes involved in breeding system control *(Gametophytic factors)*.

However, to evaluate the effect of gametophytic selection on the structure of populations, it must be considered that, if the selection in the diploid phase operates against the alleles favoured in the haploid phase, the effect is expected to be practically nil; even when the allele is neutral, the modification of gene frequencies would not be so fast. The selection is much more effective when it acts in the same direction in both haploid and diploid phases, affecting pollen and the sporophytic progeny.

The assumption is that the same genes are expressed in both the generations and confer adaptive value to both pollen and sporophyte. The result will be a very high pressure of selection and a rapid modification of the population structure, much faster than in the case of sporophytic or gametophytic selection alone. Evidence of haplo-diploid gene expression has been obtained by different approaches in many species; a quantitative estimate of genes controlling basic metabolic functions has been obtained by means of analysis of gene expression of dimeric or polimeric isozymes in maize, tomato, barley, pinus, and a very large sampling of genes has been obtained by means of mRNA-cDNA heterologous hybridization between pollen and sporophytic tissues in maize and *Tradescantia* (see Frova and Pè, this book, for a review). 60-70% (according to species and experimental method) of the genome has been shown to be shared by the gametophytic and the sporophytic phases. This includes genes involved in the control of important traits, such as pollen tube growth and reserve accumulation in the seed (*de-ga* mutants, Ottaviano et al 1988), plant vigour (Mulcahy 1971, 1974; Ottaviano et al 1982), abiotic stress resistances (Zamir et al 1982, Zamir and Gadish 1987, Searcy and Mulcahy 1985, Ottaviano et al 1990, Sari-Gorla et al 1989, 1991), phytotoxin resistance (Hodking 1988). Direct proof of haplo-diploid control of these traits has been obtained on the basis of the response to selection applied at gametophytic level that is observed in the resulting sporophytic progeny.

2. Theoretical models

Theoretical population models including gametic selection have been proposed (Harding and Tucker 1969, Harding 1975, Clegg 1978, Pfahler 1983). To make precise predictions of population structure modifications, it would be necessary to consider a complex model that took into account the different components of sporophytic fitness, the gametophytic fitness components, and the effects of interaction with the stylar tissues of the female plant. A simplified model can be constructed on the basis of w (the total adaptive value of the plant, including vitality of seed and seedling, plant vigour, fertility as pollen quantity produced and seed set) and v (pollen fitness, including microspore development effectiveness, competitive ability in terms of germinability and tube growth rate).

It is thus not difficult i) to compute the variation of gene frequencies *per* generation under sporophytic selection (SS), gametophytic selection (GS) or sporophytic and gametophytic (S+G) selection, ii) to compare the population evolution trends in the three situations, according to gene frequency and selection pressure (coefficient of selection *s*).

Gene frequency determination

Let us consider one locus having two alleles, A and a, in a random-mating population. If the frequencies of A and a are p and q, respectively, the genotypic frequencies in this population are : $p^2 : 2pq : q^2$.

If the relative adaptive values of the genotypes are: $w_{11} = 1$, $w_{12} = 1$ and $w_{22} = 1-s$, where s is the coefficient of selection, i.e. selection acts against a recessive allele, when SS is applied, the gene frequency of q (q_S) is:

$$q_S = \frac{q - q^2 s}{1 - q^2 s}$$

If the A and a pollen adaptive values are $v_1 = 1$ and $v_2 = 1-s$, respectively, when GS is applied to this population, the gene frequency of q (q_G) is:

$$q_G = \frac{q \,(1 - 1/2\, s) - q^2 \,(1/2\, s)}{1 - q\, s}$$

The gene frequency variation after both sporophytic and gametophytic selection, will thus be:

$$q_{S+G} = \frac{q_S \,(1 - 1/2\, s) - q_S^2 \,(1/2\, s)}{1 - q_S\, s}$$

In the case of selection in favour of a recessive allele, the relative adaptive values are: $w_{11} = 1 - s$, $w_{12} = 1 - s$ and $w_{11} = 1$. When SS is applied, the gene frequency of p_S is:

$$p_S = \frac{p \,(1 - s)}{1 + p^2\, s - 2p\, s}$$

Under gametophytic selection p_G is:

$$p_G = \frac{p \,(1 - 1/2\, s) - p^2 \,(1/2\, s)}{1 - p\, s}$$

and when both sporophytic and gametophytic selection are active:

$$p_{S+G} = \frac{p_S \,(1 - 1/2\, s) - p_S^2 \,(1/2\, s)}{1 - p_S\, s}$$

Population dynamics

To determine the trend of gene variation under only sporophytic, only gametophytic or sporophytic-gametophytic selection, let us assume for simplicity the same value of w and v, i.e. the same value of s in both the generations.

Table 1 indicates the number of generations required to pass from a very high frequency of q (0.9) to a very low frequency (0.0001) for a recessive allele having a deleterious effect, under SS, GS or S+G selection and with different degrees of disadvantage (s=1, 0.5, 0.1). S+G selection emerges as the most effective for all values of *s* and at all allelic frequencies. However, the comparison of the relative effectiveness of the three types of selection is particularly relevant when the frequency of q is low: in every case, even if the selection against q is severe (s=1), SS is practically unable to reduce q below 0.01, since, to pass from 0.01 to 0.0001, about 9,000 generations are required, while only 6 generations are needed to produce the same reduction on q frequency by G+S selection.

Table 2 illustrates the effects of selection in favour of a recessive, adaptive allele. Here again S+G selection is the most effective for all gene frequencies and for both values of s considered; its effect is particularly evident when one considers that, even in the case of a good advantage for the a allele (s=0.5, i.e. w_{11}, w_{12} and v_1 are 0.5, whereas w_{22} and v_2 are equal to 1), when its frequency is very low (0.0001) about 7000 generations are necessary under SS to reach appreciable levels of q, as against only 20 under G+S selection; in the case of a small advantage, no fewer than 44000 generations are needed in the first case, only 150 in the second. This means that the selection of a rare recessive adaptive allele or genic combination has a good probability of success when the selective pressure is exerted in both phases of the life cycle, whereas, when selection operates only on the plant, it is practically impossible.

TABLE 1. **Evolution of a population under sporophytic, gametophytic or sporophytic and gametophytic selection against a recessive allele.**

Coefficient of selection s:	1			0.5			0.1		
Type of selection:	S	G	S+G	S	G	S+G	S	G	S+G
Change in gene frequency									
0.9 - 0.5	1	1	0	4	6	2	29	41	17
0.5 - 0.4	0	0	0	2	1	1	9	8	4
0.4 - 0.3	3	0	1	0	0	0	12	8	5
0.3 - 0.2	2	1	0	4	2	1	20	10	7
0.2 - 0.1	5	1	1	11	3	2	58	16	12
0.1 - 0.01	90	4	4	185	8	8	924	47	43
0.01 - 0.001	900	3	3	1800	8	8	9023	45	45
0.001 - 0.0001	9000	3	3	18000	8	8	90023	45	45
Total N of generations required to pass from 0.9 to 0.0001	10000	13	12	20000	37	31	100000	220	178

Number of generations necessary for a given change in q of a deleterious recessive gene under different selection coefficients. S: selection in the sporophytic phase; G: selection in the gametophytic phase, S+G selection in both the sporophytic and the gametophytic phase.

TABLE 2. **Evolution of a population under sporophytic, gametophytic or sporophytic and gametophytic selection in favour of a recessive allele.**

Coefficient of selection s:		0.5			0.1	
Type of selection:	S	G	S+G	S	G	S+G
Change in gene frequency						
0.0001 - 0.001	6700	6	7	33000	43	42
0.001 - 0.01	233	6	5	10000	43	43
0.01 - 0.1	94	6	6	1048	45	42
0.1 - 0.2	7	2	1	54	14	12
0.2 - 0.3	4	2	1	20	11	8
0.3 - 0.4	2	1	1	12	8	8
0.4 - 0.5	1	1	0	8	8	9
0.5 - 0.9	4	7	4	29	16	6
Total N of generations required to pass from 0.0001 to 0.9	7045	31	25	44171	214	169

Number of generations necessary for a given change in q of an advantageous recessive gene under different selection coefficients. S: selection in the sporophytic phase; G: selection in the gametophytic phase, S+G selection in both the sporophytic and the gametophytic phase.

3. Consequences for population structure

Natural population evolution

On the basis of the above observations, under S+G selection a high evolution rate, in terms of allelic substitution *per* time unit is to be expected. This can be obtained without a drastic reduction in sporophytic fitness, thus with a low cost of selection. In fact, the population size is a factor limiting a high evolution rate, due to the high cost of selection: if the proportion of unfitted individuals eliminated is very high, it may result in an excessively drastic reduction of the population size, leading to the extinction of the population.

Moreover, GS has an important role in genetic load regulation, particularly with regard to complex traits; in fact, at each generation the rupture of adaptive combinations by recombination produces a large proportion of unfitted genotypes (high genetic load), which can easily be removed in the pre-zygotic phase, at a cost that can be sustained by the pollen population, due to its size, which is much larger than that of the plant population. In fact, the elimination of even a conspicuous portion of the pollen produced does not affect the size of the sporophytic generation, since the pollen that remains is sufficient in quantity to ensure normal pollination (Ottaviano and Sari-Gorla 1979).

Experimental population manipulation

The efficacy of GS can be exploited as a breeding tool. So far the major successes in the genetic improvement of crops have been the result of the application of classical breeding systems, but in recent years many new technologies have been developed, which, unlike the traditional procedures, make it possible to obtain a rapid change in plant traits. In particular, gene transfer, even from taxonomically distant species, allows the production of plants endowed with new characteristics in only one or a few generations. However, for crop species for which efficient regeneration systems are not yet available, except in the case of some specific genotypes, and especially for purposes of improvement of complex traits, this type of procedure is not suitable. Where there is genetic variability in the same species, GS offers the possibility of speeding up the traditional selection methods; it can easily be included in a conventional breeding program, since it does not require sophisticated laboratory equipment and especially because it can be applied to whatever genotype.

The possibility of using this approach for breeding purposes has been demonstrated in different species. For instance, in maize, Sari-Gorla et al (1989, 1991) submitted pollen of F_1 plants, from a herbicide tolerant and a herbicide susceptible parent, to treatment with the herbicides (a sulphonyl urea and a thiocarbamate) during pollen development or during pollen function, and used it to pollinate a female plant of recessive genotype. Pollen grains of tolerant genotype were preferentially transmitted to the progeny, which proved to be significantly more tolerant to the herbicides than the control progeny, produced by not-treated pollen.

Conclusions

Many aspects of pollen genetics, its impact on population genetic models and its practical applications remain to be explored. The main reason for this is that, although the first demonstration of differential gene transmission was given early in the century, the importance of the phenomenon continued to be underestimated until 15-20 years ago. Thus the research effort that has been dedicated to the study of pollen for theoretical and practical purposes, is not comparable to that in other fields (that of cell cultures, for instance). Moreover, there exist serious difficulties with regard to pollen fitness evaluation and the determination of the genetic control of its components. Particularly with regard to natural populations, few extensive studies have been carried out, even though competition in populations of some species has been demonstrated (see Ottaviano and Mulcahy 1989 for a review).

A very detailed analysis of selection effects on pollen competitive ability and of its consequences in the sporophytic generation has been carried out in maize (Ottaviano et al 1989,), the floral structure of which is particularly suitable for this type of study. In this species the ovules have styles of different length, increasing from the apex to the base of the ear; the basal ovules are thus fertilized by the more competitive pollen. When a mixed pollination is performed, increase of the proportion of ovules fertilized by one of the two pollen types from the apex to the base of the ear, indicates the greater competitive ability of that pollen type. This can be quantitatively measured as the regression coefficient of the proportion of kernels produced by one type of pollen as a function of their position on the ear. Selection was made only on the basis of this criterion (position on the ear), thus only for gametophytic performance, and the response to selection was observed for gametophytic (tube growth) and some sporophytic traits. Heritability of pollen tube growth rate was 0.88, meaning that, for this trait, 88% of the differences between genotypes is of genetic origin, and transmitted to the progeny, and, in particular, 69% of these differences are controlled by the haploid genome; finally, genetic variability of haploid origin released by pollen selection for sporophytic traits was 20%. These results, demonstrating that natural selection acting on pollen can produce an appreciable modification of sporophytic features, are of particular interest, since obtained in only three generation of selection applied only to the gametophytic generation

References

Clegg MT, Khaler AL, Allard LW (1978) Estimation of life cycle components of selection in an experimental plant population, Genetics 89:765-792

Frova C, Taramino G, Binelli G (1989) Heat-shock proteins during pollen development in maize. Dev Genet 10:324-332

Harding J (1975) Models for gamete competition and self-fertilization as components of natural selection in populations of higher plants. In: Gamete Competition in Plants and Animals (DL Mulcahy ed) North-Holland Publ Comp, Amsterdam, pp 243-255

Harding J, Tucker CL (1969) Quantitative studies on mating systems. III Methods for the estimation of male gametophytic selection values and differential outcrossing rates. Evolution 23:85-95

Hodking T (1988) In vitro pollen selection in *Brassica napus* L. In: Sexual Reproduction in Higher Plants (Cresti M, Gori P, Pacini E) Springer-Verlag, Berlin, pp 57-62

Mulcahy DL (1971) A correlation between gametophytic and sporophytic characteristics in *Zea mays* L. Science 171:1155-1156

Mulcahy DL (1974) Correlation between speed of pollen tube growth and seedling weigth in *Zea mays* L. Nature 249:491-492

Ottaviano E, Mulcahy DL (1989) Genetics of Angiosperm Pollen. Advances in Genetics 26:1-64

Ottaviano E, Petroni D, Pè M E (1988) Gamethophytic expression of genes controlling endosperm development in maize. Theor Appl Genet 75: 252-258

Ottaviano E, Sari-Gorla M (1979) Genetic variability of male gametophyte in maize. Pollen genotype and pollen-style interaction. Israeli-Italian joint meeting on Genetics and Breeding of Crop Plants. Monogr Genet Agr:89-106, Rome

Ottaviano E, Sari-Gorla M, Pe' E (1982) Male gametophytic selection in maize. Theor Appl Genet 63:249-254

Ottaviano E, Sari-Gorla M, Villa M (1989) Pollen competitive ability in maize. Within population variability and response to selection. Theor Appl Genet 76:601-608

Pfahler PL (1983) Comparative effectiveness of pollen genotype selection in higher plants. In: Pollen: Biology and Implications for Plant Breeding (Mulcahy DL, Ottaviano E, eds) Elsevier Sci Publ, Amsterdam

Sari-Gorla M, Ferrario S, Gianfranceschi L, Villa M (1991) Herbicide tolerance in maize. Genetics and pollen selection. In: Angiosperm Pollen and Ovules (Ottaviano E, Mulcahy DL, Sari-Gorla M, eds) Springer-Verlag, New York, in press

Sari-Gorla M, Ottaviano E, Faini D (1975) Genetic variability of the gametophytic growth rate in maize. Theor Appl Genet 46:289-294

Sari-Gorla M, Ottaviano E, Frascaroli E, Landi P (1989) Herbicide-tolerant corn by pollen selection. Sex Plant Reprod 2:65-69

Searcy KB, Mulcahy DL (1985) The parallel expression of metal tolerance in pollen and sporophytes of *Silene dioica* (L) Clairv, *S. alba* (Mill) Krause and *Mimulus guttatus* DC Theor Appl Genet 69:597-602

Zamir D, Gadish I (1987) Pollen selection for low temperature adaptation in tomato, Theor Appl Genet 74: 751-753

Zamir D, Tanksley SD, Jones AJ (1982) Haploid selection for low temperature tolerance of tomato pollen. Genetics 101:129-137

Zimmerin S, Sandler L, Nicoletti B (1970) Mechanisms of meiotic drive. Ann Rev Genet 4:409

THE ISOLATION OF SPERM CELLS, THEIR MICROINJECTION INTO THE EGG APPARATUS AND METHODS FOR STRUCTURAL ANALYSIS OF THE INJECTED CELLS

C.J. Keijzer
Department of Plant Cytology and Morphology
Agricultural University
Arboretumlaan 4
6703 BD Wageningen
The Netherlands

Introduction

This chapter deals with the study of the fertilization process in plants using micromanipulation techniques in both the experimental and the subsequent structural analytical phase of the experiments. Therefore we directly inject isolated sperm cells into egg cells and synergids. The main reasons to develop and use micromanipulation techniques are: firstly, to study the wound reaction of cells after injection with micropipettes of rather wide diameter, secondly to investigate the fate of isolated intra- and interspecific sperm cells after artificial injection into cells of the egg apparatus of intact ovules and thirdly, in case of a succesfull fertilization, obtaining interspecific hybrids. It must be mentioned that the third goal can be more easily reached by isolating not only the sperm cells (= protoplasts), but also the egg cells, followed by (protoplast-)fusion of both cells (Kranz et al.,1991). However, in the latter case the fusion product (= mainly the egg cell) has been taken out of the ovule, i.e. its natural environment for fertilization and embryogenesis, so that subsequent ultrastructural studies may reveal phenomena that are more the result of the in vitro situation than when working with intact ovules.

Important problems that we are facing in our micromanipulation work with sperm cells, egg cells and synergids, if compared with current micromanipulation work with vegetative cells or protoplasts (for some references see Toyoda et al.,1988), are: the small size of the isolated sperm cells and the well-hidden position of the cells of the egg apparatus inside the ovule. Since penetrating a plant cell wall requires micropipettes with a very sharp tip, the injection hole for the transfer of the (in this case relatively large) sperm cells cannot be located in this same pipette tip, as is the case with microinjection of liquid probes, which can be carried out with micropipettes drawn with a conventional pipette puller (Toyoda et al,1988). Therefore, either bevelled pipettes (De Laat et al.,1987) or pipettes with a real lateral hole (Keijzer et al.,1988a and fig. 1), have to be made, using a microgrinder or a microforge, respectively.

In this chapter the different steps of our artificial fertilization attempts will be outlined, from the isolation of the sperm cells up to methods for ultrastructural analysis of the injected cells.

The isolation of sperm cells

Up to now most work on the isolation of sperm cells (reviewed by Keijzer et al.,1988b and Theunis et al.,1991) was carried out in favour of their structural analysis. Thanks to these studies we learned many details about their cytological composition and behaviour outside the pollen grain or tube.

Sperm cells are mainly filled with their condensed nucleus. The small amount of cytoplasm contains different kinds of organelles, depending upon the species and the developmental stage. They generally lack a cell wall, which means that they behave like protoplasts. As a result of their origin they are surrounded by a sac of vegetative cell plasma membrane of the pollen grain. This envelope may contain one single sperm cell, in other species it may contain the two sperm cells, either connected or completely separated.

Sperm cells can be isolated from pollen tubes. In so-called tricellate species they are also present in the mature pollen grain and accordingly can be isolated from this developmental stage of the microgametophyte.

Mechanical rupture of hydrated mature tricellate pollen grains is the simplest method to obtain sperm cells (Theunis and Van Went,1989). However, tricellate species form a minority among plant species and up to now we did not use this method for our micromanipulation work.

In most cases (Theunis et al.,1991) sperm cells are isolated by osmotic bursting of in vitro grown pollen tubes, which happens a few minutes after addition of water to the (osmotically active) sugar containing pollen germination medium. Since the sperm cells are protoplasts, the osmotic pressure has to increase again by addition of (a stock solution of) the initial germination medium or any desired other medium (the so-called stabilisation medium) in order to prevent them from bursting. For our micromanipulation work they are sucked out of this medium into a lateral-hole micropipette using a micromanipulator (figs 2a-d). For this suction just a low vacuum is sufficient since the relatively large diameter of the lateral hole does not cause the high capillary suction characteristic for thin micropipettes. Liquid pollen germination media are preferred to prevent damage or contamination to the micropipette. In order to prevent undesired excessive suction into the pipette due to its capillary action at the time it touches the medium with the isolated sperm cells, the pipette has to be filled abundantly with stabilization medium in advance, or can even be used for the supply of this medium.

In *Torenia fournieri*, one of our species under study, we modified this osmotic bursting technique into a more in situ system. Therefore we slowly inject water into and through a pollinated style (fig. 3), after which the sperm cells are released osmotically from their pollen tubes and subsequently appear in the droplet that emerges at the basis of the dissected style. It will be clear that this modification can only be carried out in styles that possess a canal or loosely interconnected cells in their transmitting tissue.

The injection of the sperm cells into the embryo sac

Our first microinjections of sperm cells into embryo sacs were carried out with *Torenia fournieri* (Keijzer et al.,1988a), mainly chosen since its embryo sac emerges from the integument (Wilms and Keijzer,1985). The ovules were mechanically isolated from the placenta, transferred to a liquid CC-medium (Potrykus et al.,1979) with 20% sucrose which plasmolysed the cells of the egg apparatus, and a selected ovule (10%) clearly shows its egg cell and in this same pipette tip, as is the case with microinjection inverted microscope (figs 4a-b). Subsequently the tip of the micropipette with the isolated sperm cells in their stabilization medium (= 20% sucrose in water) was carefully pushed into the embryo sac using a micromanipulator (figs 4a-b). Since penetrating this tough wall may be difficult, penetration is preferentially carried out via the middle lamella between the two halfs of the filiform apparatus (fig. 5). For injection of either the egg cell or one of the synergids the distance between the lateral hole and the tip has to be as small as 10 - 20 /u, for injection of the central cell this may be up to 90 /u (fig. 1). The injection itself can be carried out with the same low pressure as was used for the suction of the sperm cells. However, during the previous suction of the sperm cells they must be positioned directly behind the lateral hole to prevent the transfer of an excessive volume of liquid into the receptor cell, which in our experiments frequently led to its bursting, generally via the injection wound. After the transfer of the sperm cells, the pipette has to be removed carefully and slowly in order to prevent leakage and destruction of the cell.

It must be mentioned that apart from the latter removal of the pipette, all the different actions described here have to be carried out with care and patience. In our artificial fertilization work with *Torenia* in only four cases out of 40

164

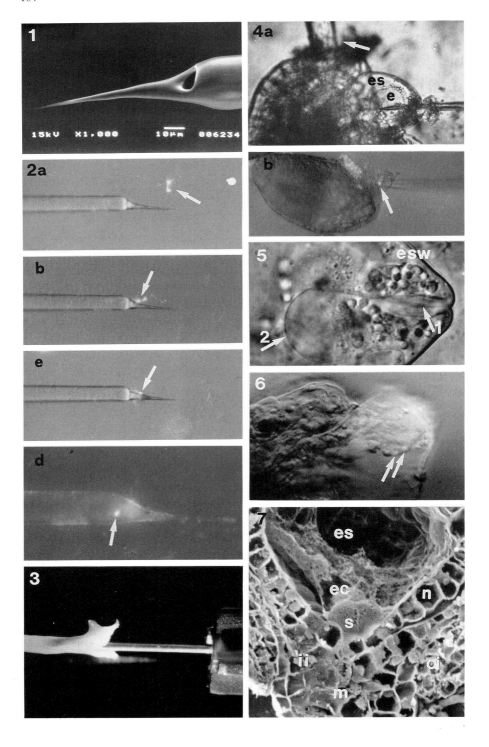

Fig.1. Pipette used for the injection of sperm cells into embryo sacs, made visible in the SEM. This is the same pipette that is used in figs 2a-d.

Fig.2a. A (pair of) sperm cells (arrow) is isolated from an osmotically burst pollen tube and is approached by the pipette that is shown in fig. 1., using a micromanipulator. The pipette is filled with stabilization medium. b-c. The sperm cells are sucked toward the pipette (arrow). d. Finally they are sucked into the pipette (arrow).

Fig.3. The in situ method to isolate sperm cells osmotically from pollen tubes inside a style of *Torenia* 10h after pollination. A conventional metal injection needle is pushed through the stigma into the stylar canal to inject water, which makes the pollen tubes burst and transfers the thus released sperm cells to the basis of the style.

Fig.4a. The injection pipette of figs 1 and 2 has been pushed into the egg cell (e) which is clearly visible inside the emerging part of the embryo sac (es). The (*Torenia*-)ovule is kept in position by microforceps, one of the beak halfs is visible (arrow). b. A situation like fig.4a with a DAPI-stained sperm cell (arrow) still in the pipette inside the embryo sac.

Fig.5. The middle lamella (arrow 1) between the two halfs of the filiform apparatus is the best site to pass the embryo sac wall (esw) with a micropipette. The wall of only one of the synergids is in focus (arrow 2).

Fig.6. Two sperm cells of *Torenia* (stained with DAPI, arrow) that have been injected into a synergid of *Torenia*.

Fig.7. Dorsi-ventral section through a critical point dried ovule of *Gasteria verrucosa*, showing the sectioned synergids (s) in front of the (intact) egg cell (ec). The embryo sac (es) is surrounded by a nucellus (n) and two integuments (ii and oi). The inner integument forms the micropyle (m).

attempts, sperm cells were succesfully transferred into the egg apparatus: twice in the egg cell and twice in a synergid (fig. 6).

Instead of microforceps also a (slightly sucking) wide-aperture holder pipette can be used. However, when proceeded isolation of the injected ovule is necessary, for example transfer to fixatives after one or two days, microforceps are preferred since they can lift the ovule out of the droplet of medium. When using a holder pipette, the ovule sometimes gets lost when leaving the droplet due to the surface tension of the latter. Secondly, sucking holder pipettes may cause undesired streaming in the medium during the microinjection work.

Although *Torenia* is well suited for micromanipulation work thanks to its emerging embryo sac, it is not representative for the large majority of plant species which hide their mature embryo sac inside at least one integument, but generally even inside a nucellus and two integuments (fig. 7). This makes vital light microscopical observation of the female target cells impossible. To overcome this problem, we recently started to develop techniques to improve the visibility of both the egg apparatus and the microtool inside it.

The visibility of the microtool inside the ovule can be improved by placing the former in even line with a 1mm-diameter light conducting glass fiber (figs 8a-b). In fact this means a modification of dark field microscopy. The more complicated the microtool used, the better it is visible due to the light dispersion in its branching substructures. Consequently, the lateral hole of our injection pipettes can be clearly seen inside the ovules. However, the visibility of its tip may be rather fainty and can be improved considerably if it is made a little dirty by touching it with a (generally fat) finger.

Making the egg apparatus visible inside the ovule may be more difficult, depending upon the plant species. In tenuinucellate or in single-integument species it may become visible using Nomarski optics. However, in crassinucellate or in double-integument species application of special detection techniques is necessary. Selective staining of the nuclei of the egg apparatus and their subsequent detection inside the intact ovule is possible by fertilization with DAPI-pre-stained pollen and observation with UV-microscopy (Keijzer et al.,1988c), which can be improved by clearing of the ovules (Willemse et al.,1990 and fig.9). Although this pollen-directed DAPI-staining is a vital process, it cannot be applicated in our micromanipulation work since fertilized ovules are not desired. Therefore experiments are in course to adapt the microtool illumination technique for UV-light, in combination with the injection of DAPI in the (roughly localised) egg apparatus using the same micropipette. In this way we might exactly localise its composing cells, which may remain vital when UV-light supply is scarce.

Also the mentioned clearing of the ovules is a lethal process and must be avoided. It probably may be turned into a vital technique when using low concentrations of glycerol. It may be avoided completely when using CLSM-microscopy with low laser- (and UV-) light intensity.

Ultrastructural analysis of the injected cells

A major problem when analysing the injected cells is to refind the small sperm cells inside the receptor cells. Up to now the best tool for ultrastructural analysis is the TEM. To avoid the time-consuming necessity to section the whole egg apparatus into ultrathin sections and their subsequent TEM-analysis, preceding analysis of semi-thin sections is necessary. Since in this way sperm cells or their fusion sites may get caught in such a section, re-embedding (Wilms, 1980) may be useful. However, recognition of sperm cells in semi-thin sections under a light microscope is difficult.

Previous localization of the sperm cells may be an important aid during the sectioning work for TEM and, as will be described later, also for SEM. Therefore the sperm cells have to be labelled with DAPI, which is added to their isolation medium. This enables their localization after injection or shortly before they are fixed for microscopical analysis. This localization can even be improved by tilting the microscopical slide about 5 for stereo photography (figs 10a-b). The consequent staining of the embryo sac cells with DAPI during the injection is not a problem in *Torenia*, given the clear observation of the artificially fertilized egg apparatus in this species. However, the DAPI/UV observation appears to damage the ultrastructure of the cells (figs 11a-b). This is even worse when clearing is used to improve the DAPI-localization.

As an intermediate step during the re-embedding technique for TEM, the semi-thin sections can also be used to make computer-assisted 3D-reconstructions. This facilitates the spatial imagination of the egg apparatus (fig. 12), an advantage that can also be obtained when using CLSM-microscopy preceding fixation.

A promising and simple technique is the sectioning or micromanipulation of critical point dried tissues preceding intracellular analysis in the SEM. Therefore, (conventionally) fixed and critical point dried tissues are stuck on a stub and the chosen cells are either sectioned using a razor blade (Cresti et al.,1986) or using a special microtome (Keijzer,1992a), followed by observation in the SEM (figs 13a and 14), or opened inside the SEM using a micromanipulator (Keijzer,1992b and figs 15a-c). Also for these SEM

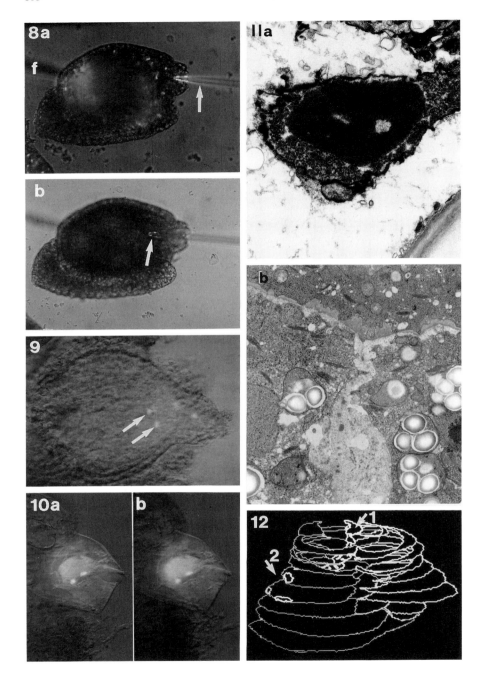

Fig.8a. An illuminated micropipette (arrow) is pushed into the micropyle of a *Gasteria verrucosa* ovule, which is kept in position between a forceps beak (f) and the microscopical slide. b. The pipette has been pushed further into the egg apparatus (arrow).

Fig.9. A cleared ovule showing sperm cells (arrow) shed from the tube of a pollen grain that had been pre-stained with DAPI during pollination. The two other fluorescent spots are presumably the vegetative nucleus (inside the micropyle) and the nucleus of the penetrated synergid.

Fig.10a-b. Stereopair (5) of photographs showing the ovule of *Torenia founieri* with two sperm cells (stained with DAPI) inside a synergid.

Fig.11a. The ultrastructure of a sperm cell inside a synergid of *Torenia fournieri* has been badly preserved due to DAPI-staining and UV-observation preceding fixation, if compared with the control (fig.b) that has been fixed without these pretreatments.

Fig.12. Three-dimensional computer reconstruction of the two synergids of *Torenia fournieri*. Inside the left synergid the pollen tube (arrow 1) can be seen, which has shed his sperm cells and the vegetative nucleus (arrow 2). The holes visible in the top part of the synergids represent the two halfs of the filiform apparatus.

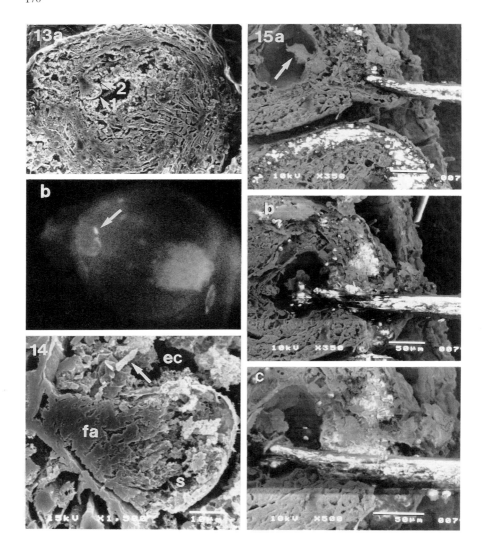

techniques preceding localization with DAPI can be useful as an aid to find the sperm cells inside the female cells (compare figs 13a and b).

Acknowledgements
The author thanks ir. PE Wittich for making the 3D-computer reconstruction, ir. VVGA Vleeshouwers for the SEM-sectioning work, S. Massalt for the photographic work and prof dr. MTM Willemse for critically reading the text.

Fig.13a. A naturally fertilized ovule of *Gasteria verrucosa* after fixation, critical point drying and sectioning, observed in the SEM. One synergid (arrow 1) is visible, as is the (damaged) egg cell (ec). b. The same ovule (at the same magnification) during its UV-observation after fertilization with DAPI-stained sperm cells. This photograph was taken before the fixation for the SEM of fig.13a. In this way one sperm cell (arrow) is localized as an aid to trace it in the SEM specimen (arrow 2). However, comparison of the two figures learns that the sperm cell has been replaced a few u's, probably due to the sectioning.

Fig.14. Detail of fig.13a. Inside the synergid (s) the filiform apparatus (fa) proceeds from the micropylar side up to halfway the cell. Inside the egg cell (ec) the sperm cell of fig. 13 is clearly visible (arrow).

Fig.15a-c. The egg apparatus (arrow) inside a dorsi-ventrally sectioned ovule of *Gasteria verrucosa* is lifted out of its position using micromanipulation inside the SEM. The charging of the specimen results from damaging its palladium-coating by the microneedle.

References.

Cresti M, CJ Keijzer, A Tiezzi, F Ciampolini and S Focardi (1986) Stigma of *Nicotiana*, ultrastructural and biochemical studies. Am J Bot 73:1713-1722

De Laat AAM and J Blaas (1987) An improved method for protoplast microinjection suitable for transfer of entire plant chromosomes. Plant Science 50:161-169

Keijzer CJ, MC Reinders and HB Leferink-ten Klooster (1988a) A micromanipulation method for artificial fertilization in *Torenia*. In: Sexual reproduction in higher plants (M Cresti, P Gori and E Pacini eds) Springer Berlin Heidelberg New York, pp. 119-124

Keijzer CJ, MC Reinders, J Janson and HB Leferink-ten Klooster (1988b) Tracing sperm cells in styles, ovaries and ovules of *Lilium longiflorum* after pollination with DAPI-stained pollen. In: Plant sperm cells as tools for biotechnology (HJ Wilms and CJ Keijzer eds) Pudoc Wageningen, pp. 149-152

Keijzer CJ, HJ Wilms and HL Mogensen (1988c) Sperm cell research: the current status and applications for plant breeding. In: Plant sperm cells as tools for biotechnology (HJ Wilms and CJ Keijzer eds) Pudoc Wageningen, pp. 3-8

Keijzer CJ (1992a) A microtome for sectioning critical point dried tissues for SEM (in preparation)

Keijzer CJ (1992b) A simple micromanipulator for SEM (in preparation)

Kranz E, J Bautor and H Lorz (1991) In vitro fertilization of single, isolated gametes of maize mediated by electrofusion. Sex Plant Reprod 4:12-16.

Potrykus I, CT Harms and H Lorz (1979) Callus formation from cell culture protoplasts of corn (*Zea mays* L.) Theor Appl Genet 54:209-214

Toyoda H, Y Matsuda, R Utsumi and S Ouchi (1988) Intranuclear microinjection for transformation of tomato callus cells. Plant Cell Reports 7:293-296

Theunis CH and JL van Went (1989) Isolation of sperm cells from *Spinacia oleracia* from mature pollen. Sex Plant Reprod 2:97-102

Theunis CH, ES Pierson and M Cresti (1991) Isolation of male and female gametes in higher plants. Sex Plant Reprod 4:145-154

Willemse MTM and CJ Keijzer (1990) Tracing pollen nuclei in the ovary and ovule of *Gasteria verrucosa* after pollination with DAPI-stained pollen. Sex Plant Reprod 3:219-224

Wilms HJ (1980) Reembedding thick epoxy sections for ultra thin sectioning. Phytomorphology 30:121-124

Wilms HJ and CJ Keijzer (1985) Cytology of pollen tube and embryo sac development as possible tools for in vitro plant (re-) production. In: Experimental manipulation of ovule tissues, their manipulation, tissue culture and physiology (GP Chapman, SH Mantell and RW Daniels eds) Longman New York London, pp. 24-36

MICROMANIPULATION AND IN VITRO FERTILIZATION WITH SINGLE POLLEN GRAINS AND ISOLATED GAMETES OF MAIZE

E. Kranz and P.T.H. Brown
Institut für Allgemeine Botanik
Universität Hamburg
Ohnhorststraße 18
D-2000 Hamburg 52

INTRODUCTION

Micromanipulation procedures are useful in the study of many aspects of plant cell biology (Schweiger et al. 1987). Examples include the possible manipulation of individual pollen grains, microincubation of pollen with substances delivered locally to the exine, pollen tube or tip of the tube by a microcapillary, and micro-injection before and after pollination. We have developed new procedures for in vitro pollen germination and describe the micromanipulation of single pollen grains of maize (*Zea mays* L.) followed by in vitro pollination and fertilization.

In vitro pollination and fertilization using excised embryo sac containing tissue and pollen, has been accomplished in various plant species (see review Zenk-teler 1990). Little is known, however, about fertilization processes at the cellular level. In vitro fertilization performed with individually selected gametes represents an effective method for studying fertilization at the single cell level and offers the possibility for examining the effects of nongametic cells from the embryo sac on the development of artificially-produced zygotes.

We report here the electrofusion-mediated in vitro fertilization of single sperm and egg cells of maize (*Zea mays* L.) and the development of the fusion products in culture as well as cell reconstruction using sperm cells and transmission of cytoplasmic organelles through the fertilization process.

1. IN VITRO GERMINATION UNDER MINERAL OIL

Experiments on the hydration state of pollen, especially of single grains, are difficult to perform. This can be circumvented by overlayering the grains with oil which prevents desiccation (Kranz and Lörz 1988, 1990; Jain and Shivanna 1990). Germination can be induced by the addition of small amounts of distilled water (in the nanoliter range) via a microcapillary to single grains previously incubated in mineral oil. Germination can also be induced when a water-delivering emulsion is used, as this emulsion provides the grains locally with the necessary water for germination in small amounts. The hydrated pollen was sprayed onto a thin layer of a moisture-delivering emulsion (Bergasol, after sun moisturiser, Chefaro 4355 Waltrop, Laboratoires Berga SA, 94152 Rungis/Paris, France; preparations of ointments Unguentum emulsificans). After the tubes had emerged, they were covered by mineral oil (Paraffin flüssig für Spektroskopie, Merck) or silicone oil (DC 200 fluid, Serva) (Kranz and Lörz 1988, 1990). Pollen grains were also germinated on a membrane of a "Millicell-CM" insert (Millipore) previously moisturized with water (Fig.1a). To prevent desiccation the membrane, together with the pollen grains, was incubated from both sides in oil. It could be shown that distilled water alone can induce the germination of maize pollen.

The sequence of events that occur when grass pollen is

on the stigma (Watanabe 1955, 1961; Heslop-Harrison 1979a,b; Dumas and Gaude 1983) - hydration, exudation, resorption and germination - could also be observed in vitro by these procedures. In many instances the pollen tubes showed oriented growth as they grew back towards the grains or when tube growth occurred along a water pathway, which might indicate hydrotropism (Fig. 1b).

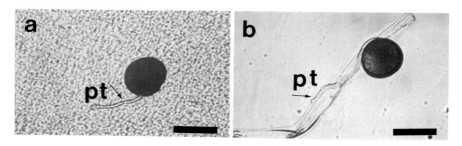

Fig. 1 a,b. In vitro germination of pollen (*Zea mays* L.) (line A 188) under mineral oil. **a** Pollen grains germinated when sprayed onto a semipermeable membrane of a "Millicell-CM" insert (pore size 0.4 μm, Millipore) which was previously moistened with distilled water. The membrane with the pollen grain was incubated from both sides in oil. **b** Oriented growth of a pollen tube along a cellulose fibre which was previously moisturized with distilled water. pt = pollen tube. *Bars* = 100 μm.

2. PENETRATION STUDIES AND MICROINJECTION OF POLLEN

The incubation of freshly pollinated silks with mineral oil did not affect the further growth of the pollen tube into the style and the emptying of grain content. The emptied grain and the tube formed a closed system having a high interior pressure of about 5000 hpa (estimated from the external injection pressures necessary to overcome internal resistance). The movement of grain contents into the style and back again into the grain could be directed by substances of different osmolarities being applied externally with a microcapillary. With solutions of up to 0.5 M sucrose an increased flow of grain contents into the style was observed. The isotonic osmotic concentration was 0.55 M sucrose (about 700 mOsm/kg H_2O). With concentrations between 0.6 and 1 M a reflux into the grain was induced.

The penetration of exogenous applied macromolecules through the pollen wall and tube was also studied. When an aqueous solution of FITC-dextran was added locally to the exine and tube in the form of small droplets, fluorescence microscopy indicated that it did not penetrate, but was concentrated at the outer walls. These observations were made on recently germinated grains and on emptied grains after pollination. Nor was there any detectable penetration of these high-molecular-weight substances, as well as ethidium bromide labelled plasmid-DNA, into ungerminated, dry but living grains or into the walls, tubes and tube tips of pregerminated grains (Kranz and Lörz 1990).

We injected dry, hydrated and pregerminated grains with FITC-dextran, Evan's blue and plasmid DNA. While grains with a low water content were easy to inject with a pressure of about 2000 hpa, pollination mostly failed. Pollination with hydrated grains, exine-injected, was occasionally successful, although these grains often dried out very rapidly when placed on the silk hairs. We used an injection pressure of about 5500-6000 hpa. Injection through the pore was more efficient than through the wall because of a lower rate of leakage and drying out of the grains after deposition onto the stigmatic hairs (Fig.2a). Pollination with pore-injected grains was most efficient and occurred at a mean frequency of about 26%.

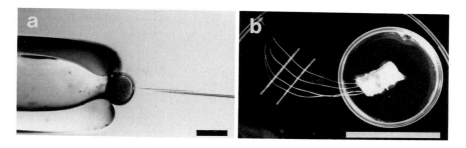

Fig.2 a,b. Micromanipulation of a pollen grain of *Zea mays* L. **a** Microinjection. *Bar* = 100 μm. **b** Arrangement of ear segment for single grain pollination under microscopic observation. *Bar* = 5.0 cm.

3. MASS AND SINGLE GRAIN POLLINATION, IN VITRO FERTILIZATION AND KERNEL DEVELOPMENT

For studies in maize, the system developed by Sladky and Havel (1976) and Gengenbach (1977a,b) allows micromanipulation and studies of pollination events under microscopic observation as well as fertilization under controlled conditions, thus avoiding the problems of regeneration and somaclonal variation induced by cell culture. Ear segments were placed in 5-cm diameter plastic dishes; these plates were then placed into 20-cm diameter dishes so that the silks hung outside the smaller dish into the larger one (Fig.2b). Single grain pollination was accomplished under microscopic observation by a thin plastic rod fixed on a micromanipulator. One day after pollination the silks were cut off and the ear pieces were cultured. Eighteen days after pollination, embryos with scutellum were removed and transferred onto germination medium (Kranz and Lörz 1990).

Raman et al. (1980) reported fertilization frequencies that increased from 4% upon single grain pollination to 50% upon pollination with ten grains. Variable fertilization rates, up to 75%, following single grain pollination were obtained by Hauptli and Williams (1988). In our experiments with single grain pollination, the mean success rate was 39% while mass pollination resulted in a mean fertilization rate of 41%. When mass and single pollination were compared, no significant differences in the success rates could be found. The plants derived from ear segment culture were shorter and flowered earlier than seed-derived plants. The mean fertilization frequency following microinjection and pollination with one grain was 29%, although rates of up to 100% could be obtained (Kranz and Lörz 1990). Although there is evidence of stoichiometric changes, no DNA polymorphism (eg. no somaclonal variation) could be found in the regenerated maize plants (Fig.3).

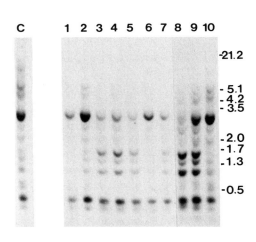

Fig.3 No significant DNA polymorphisms either between the ear segment culture derived regenerants or from the non tissue culture control (C) could be observed. Some evidence of stoichiometric differences can be seen. Genomic DNA was isolated from 10 regenerated plants (*Zea mays* L., line A 188) according to the method of Brown et al. 1991 and digested with Hind III, Southern blotted and probed with the 3Kb actin gene.

We have preliminary eviden. hat DNA can be introduced into pollen by microinjection and that these grains successfully pollinate the stigma, routinly resulting in one plant after about 3 hours microinjection of pollen grains. Work is in progress to evaluate the usefulness of pollen as a vector for transformation in maize with this method.

4. IN VITRO FERTILIZATION OF SINGLE, ISOLATED GAMETES

Isolation procedures of male and female gametes in higher plants were recently reviewed by Theunis et al. 1991. Sperm cells, egg cells and the non-gametic cells of the embryo sac of maize were isolated and individually selected. The in vitro fertilization was performed by electrofusion of defined pairs of these cells (Kranz et al. 1990, 1991a,b). Sperm cells were released from the pollen grains after rupture by osmotic shock, in the fusion medium and individually selected by microcapillaries under microscopic observation. Ovules were selected under a dissecting microscope. Ovular tissue, containing the embryo sac, was collected in mannitol solution following digestion with 1.5 ml of cell wall degradating enzyme solution followed by mechanical isolation of the cells of the embryo sacs using thin glass needles (Kranz 1991, Kranz and Lörz 1991). The yield was determined finally by the manual step of isolation (mean frequency 55%).

Single cells were individually selected and transferred by microcapillaries connected via teflon-tubing filled with mineral oil, to a computer-controlled dispenser/ dilutor (Koop and Schweiger 1985a). Fusion was performed as described previously (Koop et al. 1983, Koop and Schweiger 1985b, Schweiger et al. 1987). Ten fusion droplets of 2000 nl, consisting of 0.50 M mannitol were overlayered with mineral oil on a coverslip and selected single gametes were transferred to these droplets. Fusion was performed with a pair of electrodes which was fixed to an electrode support mounted under the condensor of the microscope. Single or multiple (2-3) negative DC-pulses (50 μs; 0.7-1.0 kV x cm^{-1}) were applied on one of the electrodes after dielectrophoretic alignment of defined gametes and cytoplasts (1MHz; 48-

71 V) with an electrofusion apparatus (CFA 400, Krüss, FRG). Fusion products were cultured on the transparent, semipermeable membrane of "Millicell-CM" dishes (diameter 12 mm, pore size 0.4 μm, Millipore). These dishes were inserted in the middle of 3-cm plastic dishes previously filled with 1.5 ml of a maize feeder-suspension. Fusion products as well as feeder-cells were cultivated in a modified MS medium (Murashige and Skoog 1962) with 1.0 mg/l 2.4 D and 0.02 mg/l kinetin.

Sperm cells fused very efficiently with other sperm cells (mean frequency of one to one-fusion 85%). The mean frequency of fusion of a sperm cell with an egg cell was 85% (Fig.4a). Fusion of an egg cell with a sperm cell without any applied electrical pulse has never been observed even with cells held in cloth contact by dielectrophoretic alignment. Isolated egg cells showed protoplasmic streaming during 22 days of culture, but no cell division could be observed. Also the fusion products of two egg cells did not divide in culture. However, under comparable culture conditions, the fusion products of a sperm cell with an egg cell developed into multicellular structures with a mean frequency of 83% (Kranz et al. 1990, 1991a,b) and into structures with roots and meristematic green spots (Kranz and Lörz 1991) (Fig. 4b,c).

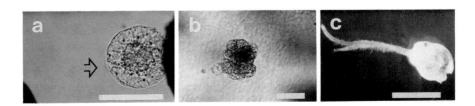

Fig. 4 a-c. Electrofusion-mediated in vitro fertilization with single gametes of *Zea mays* L. **a** Egg cell fused with a sperm cell. *Arrow*: Sperm cell inside the egg cell. *Bar* = 65 μm. **b** Multicellular structure one week after fusion on a semipermeable membrane of a "Milli-cell-CM" insert. *Bar* = 50 μm. **c** Regenerated structure 4 weeks after fusion with a root and meristematic green spots. *Bar* = 5.0 mm. Taken from Kranz and Lörz 1991. Used with permission.

The following fusion combinations could also be accomplished: Sperm + synergid (mean 75%), sperm + central cell (mean 55%), sperm + synergid with adherent egg cell and occasionally with adherent second synergid, sperm + egg cell with adherent synergids occasionally with adherent central cell, sperm + cytoplast + egg cell, occasionally with adherent synergids and central cell. The high fusion ability of the sperm cells allowed a high yielding cell reconstruction (mean 73%) (Kranz et al. 1990, 1991a,b). Fusion products of central cells and sperm cells developed in culture (Kranz and Lörz 1991).

So far techniques for the transfer of cytoplasmic characters via protoplast fusion, as well as the fusion of cytoplast and protoplast or karyoplast, to perform cell reconstitution have to date been applied to somatic cells only. Using gametes and cytoplasts of somatic cells, the possibilities for combining genetic material will be extended and will promote studies of cytoplasmic inheritance.

Recently fusion products of the gametes have been used for transformation studies. After microinjection of plasmid DNA high survival rates (mean 83%) and division rates (mean 67%) were observed (E. Kranz, unpublished). The creation of fusion products consisting of cells with "natural" competence for division and regeneration may facilitate studies of the development of artificial produced zygotes as well as studies of transformation through the sexual route.

REFERENCES

Brown PTH, Göbel E, Lörz H (1991) RFLP analysis of *Zea mays* callus cultures and their regenerated plants. Theor Appl Genet 81:227-232

Dumas C, Gaude T (1983) Stigma-pollen recognition and pollen hydration. Phytomorphology 30:191-201

Gengenbach BG (1977a) Development of maize caryopses resulting from in vitro pollination. Planta 134:91-93

Gengenbach BG (1977b) Genotypic influences on in vitro fertilization and kernel development. Crop Sci 17:489-492

Hauptli H, Williams S (1988) Maize in vitro pollination with single pollen grains. Plant Sci 58:231-237

Heslop-Harrison J (1979a) An interpretation of the hydrodynamics of pollen. Am J Bot 66:737-743

Heslop-Harrison J (1979b) Aspects of the structure, cytochemistry and germination of the pollen of rye (*Secale cereale* L.). Ann Bot [Suppl 1] 44:1-47

Jain A, Shivanna KR (1990) Storage of pollen grains of *Crotalaria retusa* in oils. Sex Plant Reprod 3:225-227

Koop HU, Dirk J, Wolff D, Schweiger HG (1983) Somatic hybridization of two selected single cells. Cell Biol Int Rep 7:1123-1128

Koop HU , Schweiger HG (1985a) Regeneration of plants from individually cultivated protoplasts using an improved microculture system. J Plant Physiol 121:245-257

Koop HU, Schweiger HG (1985b) Regeneration of plants after electrofusion of selected pairs of protoplasts. Eur J Cell Biol 39:46-49

Kranz E, Lörz H (1988) In vitro germination, pollination, micro- manipulation and transformation studies with pollen of maize. The Second International Congress of Plant Molecular Biology, Jerusalem, Nov. 13-18, Abstracts 554

Kranz E, Lörz H (1990) Micromanipulation and in vitro fertilization with single pollen grains of maize. Sex Plant Reprod 3:160-169

Kranz E, Bautor J, Lörz H (1990) In vitro fertilization of single, isolated gametes, transmission of cytoplasmic organelles and cell reconstitution of maize (*Zea mays* L.). In: Nijkamp HJJ, Van der Plas LHW, Van Aartrijk J (eds) Progress in plant cellular and molecular biology. Proceedings of the VIIth International Congress on Plant Tissue and Cell Culture, Amsterdam, The Netherlands, 24-29 June 1990, Dordrecht Boston London: Kluwer Academic Publishers, pp 252-257

Kranz E (1991) In vitro fertilization of maize-mediated by electrofusion of single gametes. In: Lindsey, K (ed) Plant tissue culture manual. Kluwer academic publishers. Dortrecht. Submitted

Kranz E, Lörz H (1991) In vitro fertilization with isolated gametes of maize and its application to study fertilization processes and early events of zygote development. In: Mulcahy DL, Ottaviano E (eds) Proc. Int. Symp. on Angiosperm Pollen and Ovules: Basic and Applied Aspects, Como, Italy, 23-27 June, Springer, New York. Submitted

Kranz E, Bautor J, Lörz H (1991a) In vitro fertilization of single, isolated gametes of maize mediated by electrofusion. Sex Plant Reprod 4:12-16

Kranz E, Bautor J, Lörz H (1991b) Electrofusion-mediated transmission of cytoplasmic organelles through the in vitro fertilization process, fusion of sperm cells with synergids and central cells, and cell reconstitution in maize. Sex Plant Reprod 4:17-21

Murashige T, Skoog E (1962) A revised medium for rapid growth and bioassays with tobacco tissue cultures. Physiol Plant 15: 473-497

Raman K, Walden DB, Greyson RI (1980) Fertilization in Zea mays by cultured gametophytes. J Hered 71: 311-314

Schweiger HG, Dirk J, Koop HU, Kranz E, Neuhaus G, Spangenberg G, Wolff D (1987) Individual selection, culture and manipulation of higher plant cells. Theor Appl Genet 73: 769-783

Sladky Z, Havel L (1976) The study of the conditions for the fertilization in vitro in maize. Biol Plant (Praha) 18: 469-472

Theunis CH, Pierson ES, Cresti M (1991) Isolation of male and female gametes in higher plants. Sex Plant Reprod 4:145-154

Watanabe K (1955) Studies on the germination of grass pollen. I. Liquid exudation of the pollen on the stigma before germination. Bot Mag (Tokyo) 68:40-44

Watanabe K (1961) Studies on the germination of grass pollen. II. Germination capacity of pollen in relation to the maturity of pollen and stigma. Bot Mag (Tokyo) 74: 131-137

Zenkteler M (1990) In vitro fertilization and wide hybridization in higher plants. Crit Rev Plant Sci 9:267-279

IN SITU AND IN VITRO ASPECTS OF EMBRYO - OVULE INTERACTIONS IN INTRA- AND INTERSPECIFIC BEET CROSSES

Lone Bruun
Botanical Laboratory
Univ. of Copenhagen
Gothersgade 140
DK-1123 Copenhagen K
Denmark

With the purpose to reestablish resistance against different diseases such as leaf spot and curly top in inbred sugar beet lines, many attempts have been made to produce interspecific crosses with wild beet species (e.g., *Beta procumbens*, Savitsky 1975). Unfortunately, the interspecific F_1 seedlings often die from root necrosis (Speckmann & De Bock 1982). On the other hand, intraspecific crosses between lines with different genetical background (e.g. different ploidy level) have successfully been made in *Beta vulgaris* (Jassem 1973), and triploids are typically used in commercial seed production. In addition, a succesful system for the production of gynogenic haploids in sugar beet (*Beta vulgaris*) through *in vitro* culture has been developed (Hosemans & Bossoutrot 1983, D'Halluin & Keimer 1986).

Although ovule culture is used for gynogenic haploid formation as well as embryo rescue, very few attempts have been made to investigate cellular aspects of changes occuring during *in vitro* culture compared to normal seed development *in situ*. Likewise, the understanding of seed development in general is significantly hampered by the lack of experimental data on how the various cell and tissue types in the ovule function in supplying nutrients to the developing seed. In this investigation growth conditions developed for haploid production (D'Halluin & Keimer 1986) have been used for *in vitro* culture of ovules from inter- and intraspecific crosses between beets with different genetical background. Structural features of *in vitro* and *in situ* grown ovules as well as zygotic and gynogenic embryos/ovules have been compared and analyzed by statistical approaches especially - by using the different physiological conditions to get some insight into the regulation of nutrient supply.

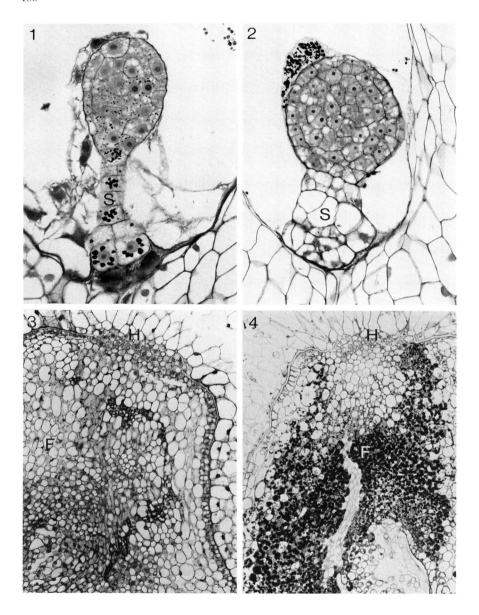

Figs 1 - 4. All sections are stained with PAS and Aniline Blue Black.
Fig. 1. Embryo from *in situ* grown ovule. The suspensor (S) contains a lot of starch grains (x 270) **Fig. 2.** Typical *in vitro* embryo. All starch has disappeared and the suspensor tissue seems more callus-like (x 380) **Fig. 3.** Funiculus (F) and hypostase (H) from *in situ* grown ovule (x 200) **Fig. 4.** Funiculus and hypostase from *in vitro* grown ovule. The funiculus tissue shows a high accumulation of tannins, starch, and protein (x 120)

187

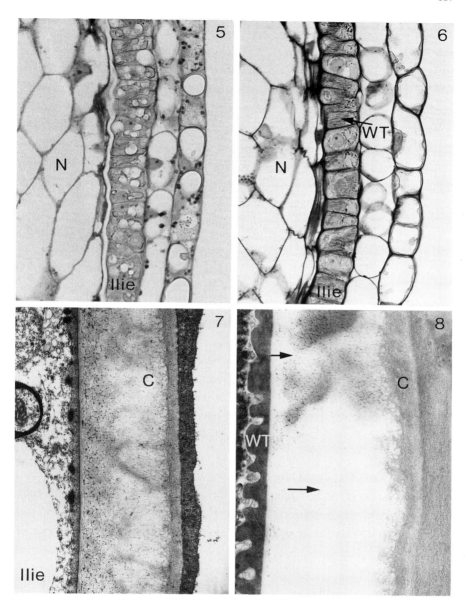

Figs 5 - 8. Figs 5 & 6 are stained with PAS and Aniline Blue Black. **Figs 7 & 8** are transmission electron micrographs.
Fig. 5. Integuments from *in situ* grown ovule(x 270)**Fig. 6.** Integuments from *in vitro* cultured ovule. Reticulate wall thickenings (WT) are developed in cell walls of the inner integument inner epidermis (II i.e.)(x 270)**Fig. 7.** Ultrathin section of the reticulate cuticle (C) between nucellus (N) and II i.e. from *in situ* grown ovule (x 15,300)**Fig. 8.** The same cuticle after *in vitro* culture. A patchy decomposition of the cutin has taken place (arrows) making transport from II i.e. to nucellus possible(x 21,200)

Material and methods

Male sterile and fertile varieties af *Beta vulgaris* L. and male fertile *B. procumbens* Chr. Sm. were supplied by Maribo Seed (Danisco A/S). Plants were grown in greenhouses and every plant from each variety was isolated in pollen-proof cabins before flowering. Ovules fixed directly from the flowering branches are referred to as *in situ* grown plant material. Ovules fixed from culture are referred to as *in vitro* grown. Methods for controlled pollination and sampling for microscopy was decribed in Bruun (1991a). For growth conditions, fixation and microscopy of the intra- and interspecific crosses see Bruun (1987,1991a) and for statistical treatments see Bruun (1991b).

Results

Suspensor. In all cross types tested, the *in situ* grown suspensor cells always contain a large amount of starch grains (Fig. 1). One of the first reactions to *in vitro* culture is a remarkable decrease in the content of starch in the suspensor cells (Fig. 2), and after a few days in culture all starch grains has already disappeared. After a longer period of *in vitro* growth the suspensor cells divide irregularly, the whole suspensor often develops a callus-like appearance (Fig. 2), and secondary starch accumulation is often observed.

Funiculus. The funiculus is sharply separated from the rest of the ovule tissue by the hypostase (Fig. 3). Compared to the *in situ* situation the funiculus cells of *in vitro* cultured ovules increase their accumulation of tannins, starch, and protein (Fig. 4). The content of tannins, starch, and protein increases by time of culture while no accumulation occurs in the funiculus tissue during *in situ* growth (Fig. 4).

Integuments. During *in situ* seed formation the integuments take part in seed coat formation. After about 5 days in culture the inner epidermis of the inner integument (II i.e.) develops reticulate wall thickenings (WT) in the periclinal as well as the anticlinal cell walls (Fig. 6). After about 14 days of *in vitro* culture 80 % of all ovules have developed wall thickenings in II i.e. whereas no wall thickenings are observed *in situ* (Fig. 5). After pollination a prominent cuticle (C) develops between II and the nucellus (Nu) (Fig. 7) and this cuticle remains intact during *in situ* seed formation. During initial phases of *in vitro* culture this cuticle shows the same appearance as *in situ* but often appears thinner. Between 3 - 7 days in culture this cuticle starts to disorganize leaving apparently cutin-free areas between areas filled up with the remaining parts of the original cuticle (Fig. 8).

Discussion

All the cellular changes mentioned here seems to be a clearcut response to *in vitro* culture since no differences were seen with respect to neither level of ploidy nor the type of crossing being analyzed. The same type of cellular responses can also be seen in anatomical reports on sugar beet ovules cultured *in vitro* for the production of gynogenic haploids (Olesen et al. 1988 and unpublished results) using the same media (D'Halluin & Keimer 1986). This strong similarity in cellular responses is noteworthy since all results in the present study were obtained on fertilized ovules whereas the gynogenic haploids are produced from unfertilized ovules.

The very fast metabolism of starch in the suspensor cells during the first days of *in vitro* culture can be interpreted as a way of keeping the embryo alive after the very abrupt cessation of the placental nutrient source upon excision of the ovule. This would support the conclusion of Raghavan (1986) that the primary function of the suspensor is to orient the embryo in close proximity to the source of nutrients in the ovule or embryo sac. Artschwager (1927) showed that the inner testa of mature sugar beet seeds has a striated appearance characterized by wall thickenings. Rapidly after transfer into *in vitro* culture young ovules of sugar beet developed similar wall thickenings (this paper and Bruun 1991a,b). Wall thickenings from the II i.e. in cotton ovules (the fringe-layer) have been decribed by Ryser et al. (1988) who concluded that the II i.e. hardly functions as an endothelium involved in nutrient transport to the developing embryo sac, but rather in transport of assimilates into the inner integument. This could proably also be the case in *in vitro* grown sugar beet ovules.

In grapefruit (Espelie et al. 1980) and in caryopses of cereals (e.g. Oparka & Gates 1982) the inner seed coat contains a continuous cuticular layer which is suggested to be a diffusion barrier controlling imbibition before seed germination. The very heavy cuticle which develops in sugar beet ovules after fertilization is believed to act as a similar barrier for the transport of nutrients and water during seed development. During *in vitro* culture this cuticle has the same general appearance as in *in situ* developed ovules, apart from being thinner, but it shows a remarkable patchy decomposition of cutin resulting in apparently cutin-free areas. This could lead to the assumption that transport of water and solutes from the inner integument to the nucellus tissue is possible during *in vitro* culture - at least in the cutin-free areas between the II i.e. and the nucellus. Thus, the cellular changes observed may well reflect a changed nutrient pathway as a response to excision and *in vitro* culture.

During *in situ* growth the funiculus represents the only possible pathway for transport of assimilates from the placenta into the developing ovule. The funiculus remains intact after excision of the ovule for *in vitro* culture. Probably the funiculus pathway of nutrients and water remains functional during *in vitro* culture in those cases were the funiculus is in contact with the substrate. The very heavy accumulation of starch and proteins in the funiculus from *in vitro* grown ovules supports this hypothesis.

In conclusion, the results of this study show that the cellular responses to *in vitro* culture of sugar beet ovules can all be interpreted as physiological responses to the sudden lack of nutrients which must occur after the abrupt cessation of the ovule from the placenta before culturing. This triggering effect apparently starts a process allowing the embryo to survive through a rapid metabolism of the suspensor starch. This metabolism is followed by the development of wall thickenings in II i.e. and a decomposition of the cuticle between II i.e. and the nucellar tissue which seems to permit the establishment of a new nutrient pathway necesarry for continued ovule and seed development during *in vitro* conditions.

The present approach of structural investigations of developing ovules *in situ* and *in vitro* demonstrates important potentials for basic understanding of the regulation of ovule development and for practical improvements in the implementation of cell and tissue culture methodology (e.g. ovule culture efficiency depending on orientation of the funiculus).

References

Artschwager W (1925) Development of flowers and seed in the sugar beet. J Agricult Res 34:1-25

Bruun L (1987) The mature embryo sac of the sugar beet, *Beta vulgaris*: A structural investigation. Nord J Bot 7: 543-551

Bruun L (1991a) Histological and semi-quantitative approaches to in vitro cellular responses of ovule, embryo and endosperm in sugar beet, *Beta vulgaris* L. Sex Plant Reprod 4: 64-72

Bruun L (1991b) A statistical analysis of some genetical, physiological and anatomical parameters of the development of in situ- and in vitro-grown ovules from intra- and interspecific crosses in the genus *Beta*. Sex Plant Reprod. 4: 118-125

D'Halluin K, Keimer B (1986) Production of haploid sugar beets (*Beta vulgaris* L.) by ovule culture. In: Horn W et al. (eds) Genetic manipulation in plant breeding. W de Gruyter & Co. Berlin New York pp 307-309

Espelie KE, Davis RW, Kolattukuty PE (1980) Composition, ultrastructure and function of the cutin- and suberin-containing layers in the leaf, fruit peel, juice-sac and inner seed coat of grapefruit (Citrus paradisi Macfed.) Planta 149: 498-511

Hosemans D, Bossoutrot D (1983) Induction of haploid plants from *in vitro* culture of unpollinated beet ovules (*Beta vulgaris* L.). Z Pflanzenzüchtg 91: 74-77

Jassem M (1973) Endosperm development in diploid, triploid and tetraploid seed of sugar beet (*Beta vulgaris* L.). Genet Pol 14: 295-303

Olesen P, Buck E, Keimer B (1988) Structure and variability of embryos, endosperm and perisperm during in vitro culture of sugar beet, *Beta vulgaris* ovules. In: Cresti M, Gori P, Pacini E (eds) Sexual reproduction in higher plants. Springer Berlin Heidelberg New York pp 107 - 112

Oparka KJ, Gates P (1982) Ultrastructure of the developing caryopsis of rice (*Oryza sativa*) in relation to its role in solute transport. Protoplasma 113: 33-43

Ragavan V (1986) Embryogenesis in angiosperms. Cambridge University Press Cambridge

Ryser U, Schorderet M, Jauch U, Meier H (1988) Ultrastructure of the "fringe-layer", the innermost epidermis of cotton seed coats. Protoplasma 147: 81-90

Savitsky H (1975) Hybridization between *Beta vulgaris* and *Beta procumbens* and transmission of nematode (*Heterodera schachtii*) resistance to sugar beet. Can J Genet Cytol 20: 177-186

Speckmann GJ, De Bock SM (1982) The production of alien monosomic additions in *Beta vulgaris* as a source for the introgression of resistance to beet root nematode (*Heterodera schachtii*) from *Beta* species of the section *Patellares*. Euphytica 31: 313-323

Stress Effects on the Male Gametophyte

F.A. Hoekstra
Department of Plant Physiology
Agricultural University Wageningen
Arboretumlaan 4
6703 BD Wageningen
The Netherlands

During the last stages of maturation and after release from the anthers, particularly, pollen dehydrates to varying extents prior to its rehydration on the stigma. Desiccation is certainly not the only one stress exerted during the life of the male gametophyte. Temperature and its interaction with cellular water activity are also important parameters in the control of vitality. In the following paragraphs these complex interactions and their practical implications will be discussed in detail.

Dehydration

At release from the anthers, pollen usually undergoes rapid drying. This involves extensive infolding of membranes, considerable decline of osmotic potential and reduction of grain volume. Protein synthesis was suggested to cease due to high solute concentrations. Respiration becomes undetectable below 20% moisture content (Fig. 1). Below 10-15% there is no free water left (Priestley, 1986) which implicates that transport processes have come to a standstill. Nevertheless, pollen from many plant species are tolerant of complete desiccation. Desiccation sensitive cells suffer from extensive membrane fusion, and leakage upon rehydration. Proline and sucrose, which are common among pollen species (Stanley & Linskens, 1974) are involved in the tolerance to high temperatures (Zhang & Croes, 1983), freezing and moderate levels of dehydration (Carpenter et al, 1987). These compounds seem to act by their preferential exclusion from the immediate vicinity of macromolecular surfaces, which amplifies the macromolecule-water interaction (Timasheff, 1982). When the hydration shell of these macromolecules is gradually lost below 25 percent moisture content (MC), proline is not an effective protectant, but sucrose is (Carpenter et al, 1987).

Much information on the behaviour of dehydrated membranes has come from studies on model membranes composed of pure phospholipids (liposomes). When the 10-12 water molecules, hydrogen bonded to the polar head group, are removed (below 25% MC), the lateral spacing of these head groups is reduced (Chapman et al, 1967). This leads to increased opportunities for van der Waals' interactions between the hydrocarbon chains of the phospholipid molecules and

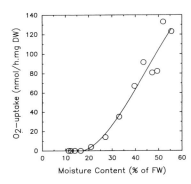

Fig. 1. Effect of water content during drying of fresh corn pollen on O_2 uptake at 25°C.

thus to formation of solid phase lipid (gel phase). During dehydration, the gel-to-liquid crystalline phase transition temperature (T_m) increases by about 70°C. Further, liposomes that are heated through their phase transition do not retain entrapped solutes (Crowe et al, 1989a). Similarly, when liposomes are prepared above their T_m and subsequently cooled through their transition, they leak. Apparently it is the phase transition, i.e. the occurrence of the two phases at the same time, which renders the liposomes transiently leaky. The question is whether such phase changes and leakage also occur in dehydrating intact cells.

Employing Fourier transform infrared spectroscopy (FTIR), Crowe et al (1989b) were able to measure phase changes of membrane phospholipids in intact pollen. An increase in T_m of about 40°C was observed during drying in desiccation tolerant cattail pollen (Fig. 2). Under dry conditions T_m just exceeds room temperature. Yet, the cells do not excessively leak solutes during rehydration when the proper precautions are observed to prevent imbibitional injury. How is this desiccation tolerance accomplished?

Desiccation tolerant organisms usually contain large amounts of di- and oligosaccharides (Crowe et al, 1984). On account of liposome experiments, it is believed now that these sugars play a unique role in the protection of membranes from the deleterious effects of desiccation (Crowe et al, 1987). Disaccharides intercalate specifically between the head groups during desiccation. Phospholipid molecules thus largely retain the original spacing between one another, and desiccation induced increase of T_m is largely circumvented. Disaccharides thus prevent leakage and fusion of dry liposomes. Reducing monosaccharides are much less effective in that respect (LM Crowe et al, 1986). In Figure 2, T_m values of dehydrating intact pollen and isolated pollen membranes (essentially free of sugars) are compared. After drying an increase in T_m of approximately 70°C was found for the membranes (and pure phospholipids), in contrast to only 40°C for membranes in the intact pollen. This difference is attributed to the presence of endogenous sucrose in the pollen (23% on a dry weight basis), the more so because addition of sucrose to the membranes depressed their dry T_m by approximately 30°C to a similar value as found for the dry intact pollen (Hoekstra et al, 1991a). The following line of reasoning may elucidate as to why sucrose effectively protects intact pollen.

When the isolated membranes of Figure 2 are dried at 20°C, they change phase at approximately 10% moisture content. Membranes in intact pollen do so at 5%. A phase change at 5% moisture content may not lead to leakage due to the lack of free water for ion transport. At 10%, however, transport may still occur to some extent, particularly because phase changes occur over a rather broad temperature range (20°C), and thus commence at a higher moisture level than 10%. It can be learned from Figure 2 that drying would have to take place preferentially at the higher temperature range, because then the chance for ion transport are minimal. Drying at 0°C may be effective to reduce metabolism and decelerate ageing, but it entails the risk of a phase change in the presence of free water, with leakage as the result.

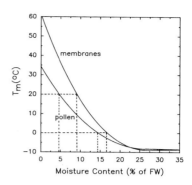

Fig. 2. Effect of moisture content on transition temperatures of membranes in intact pollen and of isolated membranes (*Typha latifolia* L.), measured by FTIR.

In some pollen species, acquisition of desiccation tolerance during development coincides with an increase in the level of sucrose, the major saccharide encountered in pollen (Hoekstra & van Roekel, 1988). Contents vary with pollen species and range from 7 to 25% on a dry weight basis. Apart from their effect on membranes, disaccharides play a role in the protection of proteins, preventing them from losing activity (Carpenter et al, 1987). The mechanism of this protection is not so clear cut as in membranes, but hydrogen bonding of OH-groups of the sugar with polar residues of the dry protein are certainly involved (Carpenter & Crowe, 1989).

Similarly as during seed development, the plant hormone, abscisic acid (ABA), occurs in immature pollen (Lipp, 1991). Desiccation tolerance of seeds is under control of ABA (Koornneef et al, 1989). ABA evokes genomic activity and production of new proteins, some of which have been linked to desiccation tolerance (Bartels et al, 1990).

Cells are more than just membranes filled with liquid. They are metabolically active entities that carry out reactions involving electron transfer. This may cause problems when water is removed, because free radicals are generated under these conditions. It is to be expected that for desiccation tolerance an effective defense mechanism against free radicals is required.

Pollen from species of *Gramineae* and *Cucurbitaceae* generally are desiccation sensitive. Apparently, protection is not sufficient in these species. It was suggested that corn pollen does not contain enough sucrose to suppress the rise of T_m with drying (Hoekstra et al, 1989). However, data of in situ T_m values were not determined. *Gramineae* pollen species are characterized by high levels of polyunsaturated linolenic acid in their phospholipids (Table I), which entail the risk of deteriorative lipid peroxidation.

Imbibitional Injury

From Figure 2 it can be learned that rehydration of dry pollen may involve a phase change of membrane lipids from gel to liquid crystalline. Unlike the sequence of events during drying, this phase change occurs in the presence of ample water. When viable dry pollen is imbibed in water of 10°C, death will ensue from the extensive leakage of solutes (Hoekstra, 1984; Hoekstra & van der Wal, 1988). Two methods are effective at preventing imbibitional injury: (a) humidifying pollen with water vapour prior to imbibition, and (b) heating pollen at imbibition. Both methods have in common that they lead to melting of gel phase phospholipid before liquid water enters the cell. Treatment with water vapour decreases T_m without the risk of leakage during the phase transition at room temperature, because it does not provide enough free water to permit the transport of ions. Imbibition of dry pollen at the higher temperature range, for instance at 30°C, results in better germination than at room temperature. The hypothesis that imbibitional injury is caused by a phase change, is further supported by data on T_m values of dry *Impatiens* pollen that is tolerant to cold imbibition. T_m hardly exceeds 0°C in this dry pollen due to its high degree of fatty acid unsaturation, so that no phase change occurs upon imbibition in cold medium, and leakage is minimal. Imbibitional (chilling) injury is a known phenomenon in all desiccation tolerant organs and organisms.

It should be emphasized that also during viability testing with such vital stains as tetrazolium salts and fluorescein diacetate, care must be taken to avoid membrane phase changes, as these stains are added to the pollen in aquaeous solutions. Prehydration of pollen in humid air and imbibition in the stain solutions at elevated temperature are meant to prevent the occurrence of false negatives.

Simon (1974) suggested that drying caused membranes of seeds and other dry organisms to assume a non-bilayer conformation, the so-called inverted hexagonal phase (H_{II}). This H_{II} phase was discovered earlier in dried brain phospholipids and is caused mainly by phosphatidylethanolamines (PEs) (see Hoekstra et al, 1991b for references). Formation of H_{II} phase is promoted by dehydration or at high

Table I. Longevity at 22°C, measured in days until half of initial viability, L_{50}, Mol percent linolenic acid in phospholipids (PL) and lipid soluble anti-oxidants, of 6 different pollen species (6% moisture content).

Species	L_{50} 22°C/6% (days)	Mol Percent 18:3 in PL	Anti-oxidant Vit E equivalents (% of DW)
Typha	150	6.1	0.11
Narcissus	83	45.7	0.75
Papaver	32	46.8	1.26
Impatiens	6	65.7	0.19
Zea[x]	0	52.5	0.12
Secale[x]	0	73.2	0.14

[x] desiccation sensitive

temperatures. The cause of leakage and death at low temperatures of imbibition was explained in terms of too slow a rearrangement of H_{II} phase lipids into the original bilayer conformation before the massive entry of liquid water takes place. Although pollen species contain considerable quantities of PEs (Hoekstra et al, 1992), ^{31}P-NMR studies did not confirm the presence of H_{II} phase lipids in pollen at 11% moisture content (Priestley & de Kruijff, 1982). Furthermore, if H_{II} phase lipid were the cause of the leakage, one would expect an aggrevation of the injury at the higher temperature range of imbibition. Precisely the opposite is generally found: an improvement of germination at elevated temperatures. Therefore, the gel-to-liquid crystalline phase change explanation of imbibitional damage is preferred over Simon's nonbilayer hypothesis.

Ageing

Lidforss (1896) made a survey of flower species, the anthers of which are either exposed to rain or protected from it. In a considerable number of species, anthers and released pollen become wet during rainy weather. Rain water is generally detrimental for pollen, as it causes loss of vitality, bursting or precocious germination. When after a short period of soaking, pollen was dried, it had generally lost its viability (Hoekstra, 1983).

Already at the beginning of this century pollen viability was known to be under the control of the endogenous moisture content and storage temperature (Pfund, 1910). Moreover, there is a strong species specific component in the longevity (Hoekstra, 1986). An example of the influence of the endogenous moisture content is given in Figure 3. Apparently ageing is considerably accelerated at elevated moisture contents. The specimen having 40% moisture content respired at an O_2-uptake rate of 1-2 ml/h.g DW. Respiration of the 15% and 6% moisture containing specimen was below the limit of detection. The desired moisture contents were established by exposure to air that is equilibrated over saturated salt solutions. Saturated NaCl, for instance, gives 75% RH at room temperature in a closed container. Dependent on the species, pollen then attains moisture contents of 14-18% (on a FW basis). We may expect that, similarly as established for seeds, longevity is doubled with every 2% reduction of the endogenous moisture content (Priestley, 1986). Similarly, longevity is doubled at every 5-6°C decrease of temperature.

Decline of viability during dry storage usually coincides with an increased leakage of solutes from the pollen at imbibition (Fig. 4). This phenomenon is

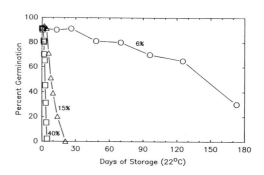

Fig. 3. Effect of endogenous water content on long term survival of *Typha* pollen at 22°C.

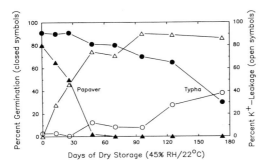

Fig. 4. Changes in viability and K⁺-leakage at imbibition during dry storage of *Typha* and *Papaver* pollen.

independent of the specific longevity of the pollen species. It suggests that reduced integrity of the plasma membrane is involved in the viability loss. At higher moisture contents (15% MC in 75% RH), the more rapid decline of viability is similarly accompanied by leakage (Fig. 5). Under both dry and semidry ageing conditions, the extent of fatty acid unsaturation of membrane lipids hardly changes. However, some new breakdown products of lipids appear, such as free fatty acids (FFAs; Fig. 5) and lysophospholipids, and the content of extractable phospholipids declines (Fig. 5). From both FFAs and lysolipids, the saturated forms, particularly, can phase separate in a membrane under dry conditions and give rise to rigid domains of high T_m, which might cause leakage at imbibition. The rise of FFAs and lysophospholipids during storage is a common phenomenon among ageing pollen species.

The curves of Figure 4 suggest that there is a strong species dependent factor involved in longevity. Table I shows that species that are proportionally rich in the unsaturated linolenic acid in their lipids tend to be short-lived. Linolenic acid (18:3) is very sensitive to free radical attack, about 4 times more sensitive as linoleic acid (18:2), that, in its turn, is approximately 40 times more sensitive as oleic acid (18:1) (Schaich, 1980). On account of this difference, one would expect a proportionally greater loss of linolenic acid from the pollen lipids than from the other fatty acids. However, the ratio of the different fatty acid hardly changes during ageing. Apparently, another mechanism of phospholipid break down than lipid peroxidation seems to operate in pollen. In seeds a mechanism of free radical induced deesterification of lipids has been proposed, which leads to loss of phospholipids and accumulation of FFAs (Senaratna & McKersie, 1986). In this respect, experiments in which pollen phospholipids were exposed to oxy free radicals are revealing. The loss of total phospholipids turned out to be dependent on both extent of unsaturation and the amount of lipid soluble antioxidants present. As indicated in Table I, the

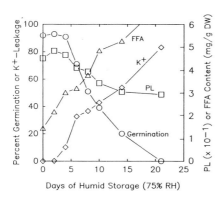

Fig. 5. Effect of humid storage of *Typha* pollen (at 15% MC) at 22°C on viability, free fatty acids (FFA), phospholipids (PL) and K⁺-leakage.

short-lived species are particularly low in endogenous lipid soluble antioxidants. Apparently, longevity of these highly unsaturated pollen species is not a critical factor during pollination, otherwise they would have been rich in antioxidants. Unsaturated pollen species also tend to perform rapid tube emergence after imbibition. The rationale behind this is unknown, but we have suggested that a high level of unsaturation and rapid tube growth are interrelated, and the result of gametophytic competition (Hoekstra, 1986).

Additional Remarks

Much of the knowledge presented here on desiccation, rehydration and ageing comes from in vitro experiments. It would be of importance to know as to how far the different stresses also influence viability in open pollinations. That drying occurs during release from the anthers and transport to the receptive stigma is evident, and most pollen species are desiccation tolerant. The amount of remaining moisture, however, will determine the rate of ageing. In a hot, humid environment when transport is delayed due to the absence of pollinators or wind, one might expect a negative effect of ageing on viability. When release from the anthers is delayed due to high humidity or rain, ageing (Hoekstra & Bruinsma, 1975; Linskens et al, 1989; Shivanna et al, 1991) or even germination (Pacini & Franchi, 1982) can occur within the anther, leading to considerable reduction of viability. Protection against ageing, for instance by extra antioxidants and less polyunsaturation of fatty acids in the lipids, may be more prominent within plant species that produce few flowers, than within species that produce new flowers on a raceme every day.

As to how far imbibitional damage occurs during pollination is unknown. Reduction of the rate of water uptake reduces the extent of the damage. This is why pollen on agar media gives better germination than in liquid media.

Imbibitional injury on stigmas may depend on the moisture content of the pollen and on the type of stigma, i.e. a dry or a wet type (Heslop-Harrison & Shivanna, 1977). My expectation is that only in the case of very dry pollen in combination with wet stigmas, viability problems might arise. Prehydration of the pollen in humid air prior to pollination solves such problems.

Desiccation tolerance offers the possibility of cold storage (below 0°C) when the pollen is dry, because ice crystals cannot be formed under these conditions. Gel phase phospholipid, however, will occur at low temperature. Thawing of the pollen and rehydration in humid air lead to melting of the gel phase, which is recommended before

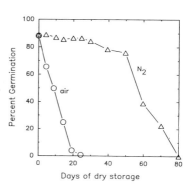

Fig. 6. Longevity of *Impatiens glandulifera* pollen during dry storage (6% MC) in air and N_2 atmosphere at 22°C.

use. Storage under oxygen free atmospheres may prolong viability if pollen is sensitive to lipid peroxidation. Data on storage in oxygen free atmospheres have been provided by Jensen (1970). *Impatiens* pollen is a good example of a rapidly ageing pollen type, having excessive amounts of the polyunsaturated linolenic acid in its lipids. When stored dry under nitrogen, longevity is prolonged several fold (Fig. 6).

References

Bartels D, Schneider K, Terstappen G, Piatkowski D, Salamini F (1990) Molecular cloning of abscisic acid-modulated genes which are induced during desiccation of the resurrection plant *Craterostigma plantagineum*. Planta **181**:27-34

Carpenter JF, Crowe LM, Crowe JH (1987) Stabilization of phosphofructo-kinase with sugars during freeze-drying: characterization of enhanced protection in the presence of divalent cations. Biochim Biophys Acta **923**:109-115

Carpenter JF, Crowe JH (1989) An infrared spectroscopic study of the interactions of carbohydrates with dried proteins. Biochem **28**:3916-3922

Chapman D, Williams RM, Ladbrooke BD (1967) Physical studies of phospholipids VI: thermotropic and lyotropic mesomorphism of some 1,2-diacyl-phosphatidyl cholines (lecithins). Chem Phys Lipids **1**:445-475

Crowe JH, Crowe LM, Chapman D (1984) Preservation of membranes in anhydrobiotic organisms: the role of trehalose. Science **223**:701-703

Crowe JH, Crowe LM, Carpenter JF, Aurell Wistrom (1987) Stabilization of dry phospholipid bilayers and proteins by sugars. Biochem J **242**:1-10

Crowe JH, Crowe LM, Hoekstra FA (1989a) Phase transitions and permeability changes in dry membranes during rehydration. J Bioenerg Biomem **21**:77-91

Crowe JH, Hoekstra FA, Crowe LM (1989b) Membrane phase transitions are responsible for imbibitional damage in dry organisms. Proc Natl Acad Sci USA **86**:520-523

Crowe LM, Womersley C, Crowe JH, Reid D, Appel L, Rudolph A (1986) Prevention of fusion and leakage in freeze-dried liposomes by carbohydrates. Biochim Biophys Acta **861**:131-140

Heslop-Harrison Y, Shivanna KR (1977) The receptive surface of the angiosperm stigma. Ann Bot **41**:1233-1258

Hoekstra FA (1983) Physiological evolution in Angiosperm pollen: possible role of pollen vigour. In: Pollen: Biology and implications for plant breeding (eds Mulcahy DL, Ottaviano E), Elsevier Publ Co NY, pp 35-41

Hoekstra FA (1984) Imbibitional chilling injury in pollen: involvement of the respiratory chain. Plant Physiol **74**:815-821

Hoekstra FA (1986) Water content in relation to stress in pollen. In: Membranes, metabolism and dry organisms (ed Leopold AC), Comstock publ Assoc Ithaca London, pp 102-122

Hoekstra FA, Bruinsma J (1975) Viability of *Compositae* pollen: germination in vitro and influences of climatic conditions during dehiscence. Z Pflanzenphysiol **76**:36-43

Hoekstra FA, Van Roekel T (1988) Desiccation tolerance of *Papaver dubium* L. pollen during its development in the anther: possible role of phospholipid composition and sucrose content. Plant Physiol **88**:626-632

Hoekstra FA, Van der Wal EG (1988) Initial moisture content and temperature of imbibition determine extent of imbibitional injury in pollen, J Plant Physiol

133:257-262

Hoekstra FA, Crowe LM, Crowe JH (1989) Differential desiccation sensitivity of corn and *Pennisetum* pollen linked to their sucrose contents. Plant Cell Environm 12:83-91

Hoekstra FA, Crowe JH, Crowe LM (1991a) Effect of sucrose on phase behavior of membranes in intact pollen of *Typha latifolia* L., as measured with Fourier Transform Infrared Spectroscopy. Plant Physiol 97: (in press)

Hoekstra FA, Crowe JH, Crowe LM (1991b) Germination and ion leakage are linked with phase transitions of membrane lipids during imbibition of *Typha latifolia* L. pollen. Physiol Plant 83: (in press)

Hoekstra FA, Crowe JH, Crowe LM, van Bilsen DGJL (1992) Membrane behaviour and stress tolerance in pollen. In: Angiosperm pollen and ovules (eds Mulcahy DL, Bergamini Mulcahy G, Ottaviano E), Springer NY 10p (in press)

Jensen CJ (1970) Some factors influencing survival of pollen on storage factors. FAO/IUFRO section 22, Working group meeting on the sexual reproduction of forest trees, Finland

Koornneef M, Hanhart CJ, Hilhorst HWM, Karssen CM (1989) In vivo inhibition of seed development and reserve protein accumulation in recombinants of abscisic acid biosynthesis and responsiveness mutants in *Arabidopsis thaliana*. Plant Physiol 90:463-469

Lidforss B (1896) Zur Biologie des Pollens. Jahrb Wiss Bot 29:1-38

Linskens HF, Ciampolini F, Cresti M (1989) Restrained dehiscence results in stressed pollen. Proc Koninklijke Nederlandse Akad Wetenschappen Series C 92:465-475

Lipp J (1991) Detection of ABA and proline in pollen. Biochem Physiol Pflanzen 187:211-216

Pacini E, Franchi GG (1982) Germination of pollen inside anthers in some non-cleistogamous species. Caryologia 35:205-215

Pfundt M (1910) Der Einfluss der Luftfeuchtigkeit auf die Lebensdauer des Blutenstaubes. Jahrb Wiss Bot 47:1-40

Priestley DA (1986) Seed aging: implications for storage and persistence in the soil. Comstock Assoc Ithaca/ London, pp 1-304

Priestley DA, De Kruijff B (1982) Phospholipid motional characteristics in a dry biological system. Plant Physiol 70: 1075-1078

Schaich KM (1980) Free radical initiation in proteins and amino acids by ionizing and ultraviolet radiations and lipid oxidation. Pt. III. Free radical transfer from oxidizing lipids. CRC Crit Rev in Food Sci & Nutr 13:189-244

Senaratna T, McKersie BD (1986) Loss of desiccation tolerance during seed germination: A free radical mechanism of injury. In: Membranes, metabolism and dry organisms (ed Leopold AC), Comstock Publ Assoc Ithaca London, pp 85-101

Shivanna KR, Linskens HF Cresti M (1991) Responses of tobacco pollen to high humidity and heat stress: viability and germinability in vitro and in vivo. Sex Plant Reprod 4:104-109

Simon EW (1974) Phospholipids and plant membrane permeability. New Phytol 73:377-420

Stanley RG, Linskens HF (1974) Pollen: Biology, Biochemistry, and Management. Springer-Verlag NY, pp 1-307

Timasheff SN (1982) Preferential interactions in protein-water-cosolvent systems. In: Biophysics of water (Franks F, Mathias SF eds), Wiley NY, pp 70-72

Zhang HQ, Croes AF (1983) Protection of pollen germination from adverse temperatures: a possible role for proline. Plant Cell Environ 6:471-476

MATURE POLLEN AND ITS IMPACT ON PLANT AND MAN

H. F. Linskens
Dipartimento di Biologia Ambientale
Università di Siena
Via P.A. Mattioli, 4
53100 Siena (Italy)

The ultimate goal of the pollen grain is to contribute to zygote formation. In principle, each mature pollen grain produced by and scattered from the anther carries the male genome twice for the double fertilization of one female embryo sac. In nature, however many more pollen grains are processed than are necessary for the fertilization of all female receptive cells, at least in most of the angiosperm plant species. This overproduction of male material is a trait that the higher plants have in common with most other plants and animals. While some people consider this a waste of energy and material, it is not so: the so-called useless pollen grains have a function with respect to other plants: in the ecosystem, they can be food for animals, and they have an impact on human beings. This impact that pollen has on plants, animals, and man is the direct consequence of its mass production.

Pollen production

Absolute pollen production is difficult to estimate (Andersen 1970). Since the beginning of this century (von Post 1919) many efforts have been made to obtain information on pollen production through calculating the surface pollen spectra. The pioneering work of Pohl (1937) was based on counting the pollen content of the stamen, with subsequent extrapolation to whole trees. But these methods included many elements of conjecture. In 1969 Erdtman summarised the then present knowledge of pollen production and dispersal. The pollen production of one Fagus tree during a 50-year period was 20.5 billion pollen grains. This was taken as a standard. On this basis he calculated a production factor for various tree species. Multiplication by the approximate weight of the individual pollen grain results in a calculation of pollen production in grams per individual tree in 1 year. We see that a beech tree produces on average about 15g pollen per year; one spruce tree, however, produces about 400 g pollen per year. Of course, some trees do not flower every year and/or with the same intensity, and most trees first begin pollen production after about 5-10 years of vegetative growth. Still we can get an indication of the order of magnitude.

Another approach starts from the number of pollen grains sedimented on a distinct surface. In a series of trials on lightships about 50 km off the coast, Hesselman (see Erdtman 1969; p. 119) collected between 8.8 and 16.2 pollen grains per square millimeter in 24 h. If we take 10 pollen grains per square millimeter per day and take an average weight per pollen grain of 30×10^{-9} g, the result is 20 mg pollen grains per day per square meter under the circumstances of an isolated sedimentation point far from vegetation.

There is another calculation for determining the contribution of pollen from *Pinus radiata*, which contains a relatively large amount of phytic acid (Jackson et al. 1982) an average from several clones of 0.13% of their weight (Jackson and Linskens 1982). The starting-point was based on a pollen production of 6 kg for one tree over 50 years; for a tree density of 770 per hectare, over a rotation of 35 years, the total pollen production would be 3200 kg/ha^{-1}, with the pollen containing 4.2 kg phytic acid ha^{-1}. If this accumulates in the top 6 cm of the forest litter layer, then pollen contributes about 70 ppm phytic acid phosphorous to this layer in a period of 35 years. This is highly significant when compared to the contribution of the needles of *Pinus radiata*, which have less than 0.0027% phytic acid. Over the same period of 35 years as much as 140000 kg needles per hectare fall to make the litter layer; nevertheless, the needles cannot contribute more than 3.8 kg phytic acid per hectare or 65 ppm phytate phosphorous. Phytic acid is the chief storage form of phosphor. We can suppose that mycorrhizal organisms can break down insoluble phytic acid salts to give inorganic phosphate capable of being utilized by plants. That means that the role of pollen in the cycling of nutrients between tree and soil has been somewhat neglected. In fact pollen is very rich in nitrogen, phosphor and other trace and essential elements (Stanley and Linskens 1974, 1984). The packaging of all these elements in the resistant pollen wall may help to preserve the nutrients for longer periods, enabling them to be more slowly released.

In general the pollen production of anemophilous plants is higher than that of entomophilic species. The ecological impact of pollen production can be surmised when one considers the fact that plantations of alder, hazel or fields of rye release an average of 10 kg pollen per year per hectare, which means that about 1 kg protein and about the same amount of lipids is delivered to the soil. One has also to take into consideration that tree pollen is only a fraction of the total pollen production in natural vegetation. The analysis of culture-influenced soil pollen spectra demonstrated that tree pollen amounts to only about 1/30 of the total pollen and spore sum (Erdman 1969). Using Pohl's (1937) figures one obtains a figure of 12.8 billion pollen grains per square meter per year in a mixed vegetation. With an average weight per pollen grain of 10 x 10^{-9} g, this means 128 gram pollen per square meter per year!

Recycling of pollen

Thus, pollen is an essential part of the recycling of organic material in an ecosystem. Packed in the resistant pollen wall, comprised of barely biodegradable sporopollenin, and loaded with a minimum of water, the pollen grains of all plant species contain very valuable substances (Linskens 1967): not only minerals, but also carbohydrates, proteins, lipids, organic acids, amino acids and enzymes, pigments, steroids and growth hormones.

During certain periods of vegetation development mass pollen shedding takes place. If this pollen falls on any water surface in vast amounts and stays drifting, it can cover the water surface with a dense, predominantly yellow layer. This is called by laymen "sulphur rain", but is in fact an indication of the first step in the recycling of pollen grains that did not match with the proper female partner.

After shedding, the pollen grains reach the leaf surface film and are often caught forever, e.g., in fissures or pores, but also in the axes of leaves or in leaf sheets, so that an accumulation of water and sedimenting particles can take place. Those niches are called phytotelmata: they have a specific community structure and are, e.g. host sites for short-lived aquatic insect communities (Maguire 1971; Frank and Lounibos 1983; Kruyt 1990). Pollen grains which land on plant surfaces and are flushed into phytotelmata deliver a valuable substrate for the exophytic microorganisms, plants, and animals. Incidental observations have shown that in such sheets of leaves of *Heracleum* pollen can accumulate and start to germinate: in most cases the grains rupture due to the low osmotic value of the aqueous substrate. Immediately there is a release of the pollen contents; as soon as the pollen grains come into contact with water, enzymes (Stanley and Linskens 1964, 1965), minerals, and amino acids (Linskens and Schrauwen 1969) diffuse out of the pollen grain. These can be used by other organisms as substrates for growth. Pollen grains are not useless at all if they do not succeed in participating in the progamic phase with the ultimate goal of fertilization! Pollen grains can also serve as food for pollinators, as orientation landmarks, and as a reward for vectors.

Pollen and soil microorganisms

When the pollen grain reaches the soil surface, either directly or washed from the plant surface, it has failed in its primary goal of participating in fertilization. In contact with a watery substrate the contents of the pollen grain is soon released, and it then becomes a substrate for the microflora of the soil. There are fungi which are specialized to subsist on pollen, e.g., the phycomycete *Rhizophidium*, which parasitizes pollen and can be regularly found on pollen precipitation. The hyphae of this fungus penetrate the pollen wall by means of haustoria, which then branch into the inner space of the pollen and form on the surface of the grain zoosporangia. After a short time the zoosporangia release zoospores ready to attach to other grains. Their special enzyme complement enables them to dissolve the exine and to enter the living pollen, using its reserve material as substrate. Other pollen is consumed by lower animals, which consume a large number of pollen grains. In some collembola (e.g., *Onychiurus pseudofimetarus*) the digestion process can be followed through the transparent exoskeleton after ingestion. Enzymes in the intestine, called exinases, are able to break down the resistant sporopollenin of the pollen exine. Bees, beetles, and most other insects, however, excrete the exine after digesting the contents of the pollen grain, including the protoplasm, since they lack the degrading capacity for the exine. The evolution of exine-degrading enzymes has apparently occurred only rarely in the animal kingdom. The obligate specificity of such aerobe fungi for a pollen germination substrate and the fact that deep pollen deposits do not accumulate in large quantities, which year after year are deposited in the soil, suggest that in the case of the insect collembola, the exine degraditive enzymes must have evolved in the soil microflora at several times.

Pollen as phytopathological amplifier

Pollen sediments not only on the soil, but also on the leaf surfaces of plants. This ecological niche, called the phyllosphere, is also the site of attack for saprophytic microorganisms, resulting in infection processes and diseased plants. The saprophytic phyllosphere is the site of interaction between fungal spores and their germination and colonization on one side, and of pollen sedimentation on the other side. The first indication of an interaction between pollen and fungal growth came from Schopfer (1934), who observed that anther extract enhances the growth of Phycomyces in a manner similar to that of yeast extracts. Incidental observations in the 1960s (Ogawa and English 1960; Bachelder and Orton 1963) suggested that pollen grains or their metabolic products have an effect on the infection process of certain fungi.

Intensive experiments have been carried out, and these have demonstrated the definitive influence of pollen grains sedimented on the leaf surface on the infection process (Chou-Chou and Preese 1968; Chou-Chou 1970; Fokkema 168, 1971). Not only vegetative leaves, but also petals and fruit of strawberry were the subject of enhanced infection in the presence of pollen grains. In the experiments it could be shown that the addition of a pollen suspension or of dry pollen grains to the leaf surface, which simulates the situation in the field, had an amplifying, stimulating effect on infection. The addition of pollen to fungal spores (*Helminthosporium sativum*) displaying a low germination rate on the surfaces of leaves substantially enhanced the percentage of germination, and also led to an increase in the number of germ tubes per surface unit (Fokkema 1971).

There are three different explanations for the effect of pollen in amplifying the infection process:

(a) The wettability of the leaves can be changed by the presence of pollen grains, e.g., by a reduction in the water repellency of the leaf surface by the cutinase secreted by the pollen grains. Cutinase secretion, which is responsible for the biodegradation of the cutin layer, has been demonstrated in germinating pollen (Linskens and Heinen 1962). In this way the pollen can directly affect the wetness of the leaf surface, and indirectly the development of the phyllosphere flora and thus the infection process.

(b) A second possibility is that the presence of pollen on the leaf surface might stimulate the production of toxins by the pathogenic fungus and thereby induce the ultimate necrosis of the leaves. It has been postulated that a certain factor, called "aggressin" (Chou-Chou 1970), may be responsible for the expansion of lesions on the leaves.

(c) The third possibility is that certain substances, secreted actively or released passively, have an effect on the pH of the germination substrate of the inoculating pathogen spores or contribute nutritively to the germinating and growing fungus. It is a well-known fact that leaf surface microorganisms have antagonistic effects (Van den Heuvel 1970). In addition to antibiotic substances from pollen, nutritive compounds, of which the pollen grains of many plant species are rich in, can promote infection. Natural pollen deposit on a hosts surface may act by delivering nutrients, that are directly useful to the pathogenic fungus. Of course, the ultimate effect of pollen grains in the phyllosphere will depend on the number of pathogenic units, the ratio of the number of pathogenic and saprophytic organisms on the leaf's surface, and also on the number and physiological activity of the pollen grains that can interact, either directly or indirectly with the infection process.

Pollen as virus carrier

Generally an abnormal morphology of pollen grains is considered to be the result of chromosomal and breeding irregularities. But in addition to hybridization, meiotic aberrations, and/or apomixis virus infection can produce abnormal pollen grains: they can be collapsed and grooved and show sunken opercula or incomplete development (Height and Gibbs 1983; Franki and Miles 1985; King and Robinson 1967; Sharma and Dubey 1982). Sometimes even fused grains and grains with "satellites" have been observed (Johnson 1989). The viruses can be carried in the inner space of the pollen grain, (Huang et al. 1983) but the outersurface can also be contaminated (Hamilton et al 1977). These facts are important ones to be taken into consideration when pollen grains are used in biotechnological approaches as carriers for gene transformations (Shukla et al 1991) or when pollen tubes are used to deliver external DNA (Lou and Wu 1988).

Pollen as feed for animals

For many insects, especially bees, pollen is a principal source of normal food, at least for part of their life cycle, because it contains the broad spectrum of essential nutritional elements for the production of royal jelly, the nourishment of larval queens and young worker larvae. Pollen is the key source of protein and lipids for larvae and imagines of all species and genera of Apidae. The amount of protein and fat in nectar, the other essential food source, is insignificant. Worker bees use the protein directly from the ingested pollen; queens, imagos, larval queens, and young larvae of both sexes receive their protein by the royal jelly produced by nurse bees supplied with pollen. To guarantee an optimal supply of pollen which is essential for the normal growth and development of the individual bee as well as for the well-being of the colonies on a whole, pollen is collected and stored in the combs of the hive. The transformation the pollen undergoes during storage results in the formation of the so-called bee-bread.

Some flower-visiting animals ingest pollen only incidentally, e.g., suspended in nectar or water. Nothing is known about importance of the contribution of pollen to animal feed with respect to vitamins, essential amino acids and hormones. Surely some animals that function as vectors in the pollination process also get pollen as a reward for their involuntary efforts in bringing the sexes together.

The great majority of the pollinators are bees and flies (Petanodou and Vokou 1990). By the way: Honey bees are poor pollinators (Westerkamp 1991), even if they are considered the queens of pollinators. Visits to flowers result in pollination only if pollen is transfered inadvertently because the visits are aimed at pollen harvesting!

Interestingly, the energy delivered by pollen to the pollinators is higher than that which is stored by wind-pollinated plants in their pollen grains (Heinrich and Raven 1972, Collin and Jones 1980).

Man and pollen

Being surrounded by clouds of pollen grains and spores all his life human beings have a manifold relationship with pollen. Besides the direct effect that pollen has on his health, man can also use it as an accessory food. In addition, pollen can play a role in human civilization in many ways.

Pollen gave rise to a separate branch of biology, called palynology, which uses fossil pollen grains as micro-fossils for the investigation of prehistoric vegetation, climate, and agriculture. The visual representation of pollen grains is used to construct pollen diagrams (palynograms): from the relative frequencies of various tree, herb, and grass pollen in fossil and subfossil sediments (biostratigraphy), conclusions can be drawn with respect to the occurrence, distribution and migration of plants in prehistorical times (historical plant geography). In recent years those palynological studies have had a growing significance on economic geology. Palaeobotany has acquired a powerful method in palynology, and pollen grains have been recognized as microfossils together with algae (diatoms, dinoflagellatae, chitinozoa, foraminiferae).

Besides being used as a tool in plant breeding (Linskens 1987), the pollen-man relationship can be divided into direct effects on man, in the form of food and the unpleasant experience of allergy, and the indirect and scientific importance of pollen, demonstrated in pollen analysis, melitopalynology, pharmacopalynology, forensic palynology, kopropalynology, and aeropalynology. We will consider some of the special aspects of pollen application on human civilization to demonstrate the important impact of pollen on man.

Human consumption

Pollen is actively marketed as a beneficial diatary supplement. The present rage for "natural food" has promoted the consumption of pollen, mostly bee-collected material, alone or in combination with other compounds, as e.g., honey or royal jelly, fermented or infermandet. The traditional school medicine is reserved about the special effects of a pollen diet. However, one thing can be said: there are no reports that pollen ingested even in large amounts has any harmful effect.

There are many reports on the curative effects of a pollen supplemented diet, e.g., in chronic prostatis, pollen is found to have a favorable influence. Also, a bleending stomach and respiratory may be reduced by the oral application of pollen or pollen extracts.

Pollen allergy (iatropalynology)

Air constitues 80% by weight of the daily intake of a human being by his 22,000 breaths a day. The main constituent of the airborne particulate metter, even in clean air, which is entering the human body, are pollen grains. The omnipresence of pollen and spores in the respiratory breath was detected and first described by Blackley 1873. Airborne pollen grains are the bio-pollutants that cause human allergy (Knox 1979). No wonder that pollen has attracted the attention of aerobiologists, meteorologists and allergists for many years. On the basis of many observations a spectrum of airborne particles can be distinguished for a certain locality. Such spectra of airborne pollen have been prepared for many localities in every continent of the world, and documented in pollen picture books. They form the basis of intensive attempts to design pollen prediction calendars. But the number of pollen species in every pollen spectrum varies, not only from place to place, but also from time to time, even from year to year. So while it is possible to demarcate certain aerobiological zones with respect to possible asthmatic attacks, the value of those pollen spectra is only retrospective, and therefore relative. A new approach is the inclusion of phenological data, so that biological parameters can be included, and not only the variation in weather conditions (Driessen 1991).

There can be no doubt that allergies against pollen have been know about for more than 100 years, and at present it is a major branch of medical allergy research. The first description of a pollinosis or hay fever was given by the american physician John Bostock in 1819. But it was the English medical doctor Charles Blackley, who suffered himself from pollinosis, who solved the fundamental problems of pollen allergy by intensive research on the causes and nature of *Catarrhus aestivus*, also called hay-asthma. He introduced pollen traps, made long-term observations on pollen deposition and pollen content of the air, and made exposition and skin tests. He found that pollen grains start to germinate after contact with the moist mucosa: "...This development of the pollen tube may be seen to occur in a very small number of cells, when placed under the microscope... and there can be very little doubt that the same change takes place, when pollen is brought into contact with the mucous membranes of the respiratory passages" (Blackley 1873).

The allergenic substances which are carried by many pollen species are liberated as soon as the pollen grain reaches a watery substrate: the allergenic substances are water soluble. Thus, the pollen grain is the carrier of the allergen(s), and not the allergen itself (Jorde 1972, Jorde and Linskens 1978, 1979; Linskens 1979). Pollen grains of many plant species (Duchaine 1959) are known to be carriers of allergens. There is some evidence that some allergens are located in the exine of the pollen grains and can diffuse out of it as early as a few minutes after contact so that the sensivite patient gets the allergenic reaction quickly. The target organs are the mucosa of the nose, the eyes, and the skin. In very sensitive patients fewer than five pollen grains are sufficient to generate an allergic attack.

Allergens are antigens, which in human beings cause certain immunological reactions. The substance with allergenic effects has to have the following properties (Wittig et al. 1970):

(1) it has to be a compound foreign to the human body,

(2) it has a molecular weight of more than 10.000 daltons,

(3) the molecular structure has a certain stability, which may be caused by aromatic groups, disulfide bridges or double bonds,

(4) the molecules have polar terminal groups as antigen determinants, which react antibody specific,

(5) the molecule will be metabolized after a certain time in the human body.

Most of the pollen allergens which have been isolated are glycoproteins. On the whole they are pollen species specific, but they may have a certain degree of crossreactivity.

In addition to being localized on the outside of the pollen grains, the submicroscopical spaces of the exine, allergens are also localized in the inner space of the pollen grain. Pollen diffusates contain a considerable amount of protein fractions. Besides those from the exine cavities, which have been added to the grains during the final development by the tapetal cells within the anthers, others originate from the inner intine. These are derived from the pollen protoplast. There is the possibility that different forms of allergens are present in the different sites of the grain.

Most interesting is the re-discovery of the amyloplast as a site of allergens by an Australian group (Singh et al. 1990), a fact that had already been described by Wolff-Eisner in 1906! They confirmed that the major allergenic protein from rye grass pollen is located in the starch granules. When a pollen grain encounters water, it bursts and releases 1,000 and more starch granules from each pollen grain into the environment. This means an amplification of the allergenic molecules in the environment, and may also explain why patients can get allergenic reactions at times of low pollen counts in the atmosphere. Also why maxima of pollen release by allergenic plants have not to be coincident with maxima of allergic attacks of patients in the respective area. As soon as the pollen grains rupture due to the presence of water, e.g., in the atmosphere, in rain clouds or in fog, the released starch grains transfer the allergenic protein, which function as carriers on its own. Furthermore it cannot be excluded that also pollen allergen molecules are detached from their carriers and reach sensitive patients as free floating allergens, passing through filters of air conditioning installations and do their job in causing the disorder characterized by wheezing, coughing, sneezing, difficulties in breathing, reddishing of the skin, and a suffocating feeling, the syndrome called hay-fever or pollen asthma.

Forensic palynology

In our human society solving criminal cases is part of the public security. In some cases pollen analysis has been used in criminal technology. The season-dependant release of pollen species has been used in indirect evidence judgements. Also, the local origin of objects in relation to criminal facts can be indicated by attached pollen grains. In combination with mineralogical and soil analysis the analysis of spectra of mature pollen can be used for confirmation of suspicion.

Kopropalynology

In the same direction is the application of the analysis of pollen spectra in faeces of man and animals. Due to the chemical properties of the sporopollenin pollen grains are extremely resistant to enzymatic degradation of the vertebrate stomach and intestine, so that they can be identified after excretion. Conclusions can be drawn with respect to food and feed used. Also, the feeding habits of animals like sheep, donkies, and elephants have been analyzed using pollen residues in faeces. The flowers visited by certain bat species have actually been determined by the pollen grains found in their excretions. Even fossil guano has been analyzed; their pollen analysis indicated the vegetation in the region during the period of deposition. Koproliths, fossilized particles of faeces in fossil animals, have indicated the feeding habits, the vegetation, and the climate during the lifetime of the fossilized individuals.

Pharmacopalynology

Mature pollen grains and their characteristic structures are used in drug analysis in pharmacy. Recognition and falsification of medical plants are sustained by analysis of the pollen grain species found in the preparations.

Melitopalynology

The pollen grains found in honey indicate the geographic origin of the collected nectar. Pollen can fall into the nectar in the flower in situ, but it can also be drawn by the nectar-collecting worker bee into the honey sac. In the latter case the pollen grain will accompany the nectar during its processing in the honey-bee's stomach. The odds of the pollen grains falling into the nectar in situ in the flower or alternatively being picked up by the hairs of the collecting bee is likely to be different for different flower types. Furthermore, some pollen is removed from the nectar while in the honey sac. Cracked pollen grains have never been found in the honey-bee's stomach. The pollen is mechanically transported from the mouth to the stomach by a slight peristaltic action along the oesophagus; it seldom stays longer than 20 minutes in the stomach. From there it is rapidly transferred to the honey stopper of the mid-intestine. The four pro-ventriculus valvae, each with a fringe of hairs, retain the pollen grains; they are compacted into a bolus that is surrounded by the peritrophic membrane. This has a network-like structure, but changes with the type of food. It tightens around the pollen mass and forms a saussage-like tube in which the pollen passes the oesophagus invagination and the ventriculus. The surrounding peritrophic membrane protects the bee's digestive tract from the sharp spines and abrasive exine surface of the pollen. Usually 1 to 3 hours are necessary for the passage of the pollen package through the ventriculus. During this time it undergoes changes: an extraction of the nutrients, after germination-like swelling, mostly in the pore area of the grains, takes place. The cellulosic and pectic material as well as the sporopollenin coat are generally indigestible to the insect. Digestion progresses rapidly and removes most of the cell's contents. Enzymatic activity of the digestion process is highest in the front midgut, whereas the absorption of metabolites from the pollen is highest in the lower end of the ventriculus (Stanley and Linskens 1985; Cowen 1988).

Pollen is the most important requirement for the bee growth, especially in feeding broods. In fact, the greater part of all nitrogen is derived from pollen protein.

The nutritive value of pollen varies with the pollen-delivering species. Three different pollen food types can be distinguished, the amount of soluble proteins being the primary discriminating factor. In general, entomophilic pollen has a higher nutritive value than anemophilic pollen. Sometimes substances toxic to the bees (anemonin from *Ranunculus* species; saponines in *Aesculus* and

Tilia; the highly toxic galitoxin in *Asclepias*) have been found in pollen, but there is only a slight chance of contamination in stored honey. While bees may be resistant, the honey may be dangerous for human consumption. In India honey heavily contamined with *Lasiosiphon* pollen has caused severe nausea and vomiting.

The identification of pollen species in honey is an important tool for the control of honey production: its origin and the way it has been handled during its extraction from the combs by the apiarists is especially important (Straka 1975, p. 189-203).

Melitopalynology is based on a detailed knowledge of special pollen morphology (Maurizio and Louveaux 1967).

Pollen in the air (aeropalynology)

Pollen grains are an essential part of the aeroplankton, the total of powdered matter's floating particles in the air. The pollen grains, mostly from anemophilic flowers, are actively or passively released from flowers and inflorescences into the surrounding air. Distribution within the air depends not only on the properties of the pollen grains - their absolute and relative weight, sedimentation velocity, and morphology - but also, and essentially, on the dynamic conditions of the air. Horizontal air streams are responsible for the distance of transport. Vertical air streams carry pollen upward in the atmosphere during the daytime, sometimes to very great altitu. The sedimentation speed of the individual pollen grain is a decisive factor. In general, sedimentation speed is lower for anemophilic pollen than for entomophilic species, with a lot of exceptions. Slow sedimentation provides an opportunity for the pollen grain to stay longer time in the atmosphere and therefore to be transported farther. The presence of pollen in the atmosphere has been confirmed by collection during aeroplane flights (Rempe 1937) and by collection d gas balloon travel and floating (Linskens 1987; Linskens and Jorde 1986). Up to 1,000 meter above the ground the concentration of pollen grains decreases only slightly, with a maximum in the range of 200 to 500 meters above the ground, always depending on the general weather conditions. The selective factor for the discriminating sedimentation is more the sedimentation velocity during the night time rather than the uplifting during the day.

Pollen in the air has two impacts on man. First, it means that all men are surrounded by pollen all through their life-time and everywhere. The concentration of the surrounding clouds of pollen grains is variable in time and space of course. In certain periods of the year up to 100,000 pollen grains per cubic meter are present for shorter or longer periods of time; it is a suspension in which individuals are dipped in. The natural consequence is that sensitive persons can suffer from a pollen allergy depending on their specific sensitivity and the amount of a specific pollen grain species present at the time. Secondly, the air body transported by the dynamics of the atmosphere can be characterized by the pollen spectrum carried with it. The pollen grains can be considered to be little balloons just tracing air movement. It turned out that the Antarctic region around the South Pole is a sort of sink for air-floating pollen. Generally speaking: pollen grains are everywhere in the atmosphere, even above the oceans and in the most remote areas of the polar caps (Linskens et al. 1991). The long-distance transport of pollen is firmly established. Given the special load the air masses get above continents as a result of the special regional flora plants, it may be possible to follow the long-distance transportation of air and trace it by the by-load of pollen grains.

Conclusion
Mature pollen has in addition to its crucial role in plant fertilization an important impact on the life and science of man.

Literature cited

Andersen S Th (1970) The relative pollen productivity and pollen representation of north european trees, and correlation factors for tree pollen spectra determined by surface pollen analyses from forests. Dan Geol Undersgelse II R No 96

Bachelder S, Orton ER (1963) *Botrytis* inflorescence blight on american holly in New Jersey. Plant Dis Rep 46: 320

Blackley CH (1959) Experimental researches on the causes and nature of *Catarrhus aestivus* (hay-fever or hay-asthma). Baillière Tindall & Co London 1873. Reprint by Dowsons of Pall Mall, London, 1959

Chou-Chou MML (1970) Biological interactions on the host surface influencing infection by *Botrytis cinerea* and other fungi with pollen grains. Thesis Univers Leeds

Chou-Chou M, Preece TF (1968) The effect of pollen grains on the infection of *Botrytis cinerea*. Ann Appl Biol 62: 11-22

Collin LJ, Jones CE (1980) Pollen energetics and pollination modes. Am J Bot 67: 210-215

Cowan JW (1988) Pollen in honey. Plants to-day 1: 95-99

Driessen MNBM (1991) Pollen and pollinosis. Medical and botanical aspects. Thesis Katholieke Universiteit Nijmegen

Duchaine J (1959) In: JM Jamar (ed) International Textbook of allergy. Thomas Springfield p. 154

Erdman G (1969) Handbook of palynology. Munksgaard Copenhagen

Fokkema NJ (1968) The influence of pollen on the development of *Cladosporium herbarum* in the phyllosphere of rye. Neth J Plant Pathol 74: 159-165

Fokkema NJ (1971) The effect of pollen in the phyllosphere of rye on colonization by saprophytic fungi and on infection of *Helminthosporium sativum* and other leaf pathogens. Thesis University of Amsterdam

Franck JH, Lounibos LP (1983) Phytotelmata: terrestrial plants as host for aquatic insect communities. Plexus Publ Medford

Franki RIB, Miles R (1985) Mechanical transmission of sowbean mosaic virus carried on pollen infected plants. Plant Pathol 34: 11-19

Haight E, Gibbs A (1983) Effects of viruses on pollen morphology.Plant Pathol 32: 369-372

Hamilton RI, Leung E, Nichols C (1977) Surface contaminations of pollen by plant viruses. Phytopathology 67: 395-399

Heinrich B, Raven PH (1972) Energetics and pollination ecology. Science 176: 597-602

van den Heuvel J (1970) Antagonist effects of epiphytic microorganisms on infection of dwarf bean leaves by *Alternaria zinnae*. Thesis Univers Utrecht

Huang B, Hills GJ, Sunderland N (1983) Virus-infected pollen grains in *Paeonia emodi*. J exper Bot 34: 1392-1398

Jackson JF, Jones G, Linskens HF (1982) Phytic acid in pollen. Phytochemistry 21: 1255-1258

Jackson JF, Linskens HF (1982) Conifer pollen contains phytate and could be a major source of phytate phosphorous in forest soils. Aust For Res 12: 11-18

Johnson MF (1989) Pollen morphology from virus-infected *Ageratum houstonianum*. Bot Gaz 150: 63-67

Jorde W (1972) Warum muß man zwischen Allergen und Allergenträger unterscheiden. Schriftenreihe Allergopharma 2: 51-54

Jorde W, Linskens HF (1978) Pollen als Allergenträger. Allergologie 1: 7-10

King RM, Robinson H (1967) Multiple pollen grains in two species of the genus Stevia. Sida 3: 165-169

Knox RB (1979) Pollen and allergy. Studies in Biology no. 107 Arnold

Kruyt W (1990) Fytotelmata, baden voor kleine fauna. Biovisie 70: 149

Linskens HF (1967) Pollen. Hb Pflanzenphysiol 18: 368-406

Linskens HF (1979) Pollen und Pollenallergie. Allergologie 2: 210-211

Linskens HF(1982) Pollen collection during a balloon trip. Incompatibility Newslett 14: 116-120

Linskens HF (1987) Pollen as a tool of the plant breeder. Biol Zbl 106: 3-11

Linskens HF, Bargagli R, Focardi S, Cresti M (1991) Antarctic moss turf as pollen trap. Proc Kon Nederl Akad Wet (Amsterdam) 94: 233-241

Linskens HF, Heinen W (1962) Cutinase in Pollen. Z Bot 50: 338-347

Linskens HF, Jorde W (1986) Pollentransport in großen Höhen - Beobachtungen während der Fahrt mit einem Gasballon. Allergologie 9: 55-58

Linskens HF, Schrauwen J (1969) The release of free amino acids out of the germinating pollen. Acta Bot Neerl 18: 605-614

Luo ZX, Wu R (1988) A simple method for the transformation of rice via pollen tube pathway. Plant Mol Biol Rep 6: 174-185

Maguire B Jr (1971) Phytotelmata: biota and community structure determination in plant-hold waters. Annu Rev Ecol System 2: 439-464

Maurizio A, Louveaux J (1967) Les méthodes et la terminologie en mélissopalynologie. Rev Palaeobot Palynol 3: 291-295

Ogawa JM, English H (1960) Blossom blight and green fruit rot of almond, apricot, and plum caused by *Botrytis cinerea*. Plant Dis Rep 44: 265-268

Petanidou T, Volou D (1990) Pollination and pollen energetics in mediterranean ecosystems. Am J Bot 77: 982-992

Pohl F (1937) Die Pollenerzeugung der Windblütler. Beih Bot Zbl 56: 1, 365-470

Von Post L (1919) Skogzradpollen i sydsvenska torvmosselagerfoljder. Forh 16 Skand Naturforskermöte Kristiania 433-465

Rempe H (1937) Untersuchungen über die Verbreitung des Blütenstaubes durch die Luftströmungen. Planta 27: 93-143

Schopfer WH (1934) Sur l'existence dans les pollinies d'Orchidées d'un facteur de croissance de microorganisme. C rend Séanc Soc Phys Hist nat Genève 51: 29-30

Sharman I, Dubey GS (1982) Histological changes in urdbean flowers due to urdbean leaf crinkle virus infection. Z Pflanzenkrankh Pflanzensch 90: 63-67

Shukla DD, Gough KH, Xiaowen X, Frenkel MJ, Ward CW (1991) Genetically engineered resistance in plants against viral infection. In : J Prakash RLM Pierik (eds) Horticulture - New technologies and application p. 107-113, Kluwer, Dordrecht

Singh MB, Hough T, Theerakulpisut P, Avjioglu A, Davies S, Smith P, Taylor P, Simpson RJ, Ward LD, McCluskey J, Puy R, Knox RB (1991) Isolation of cDNA encoding a newly identified major allergenic protein of rye-grass pollen: Intra-cellular targeting to the amyloplast. Proc Natl Acad Sci USA 88: 1384-1388

Stanley RG, Linskens HF (1964) Enzyme activation in germinating *Petunia* pollen. Nature 203: 542-544

Stanley RG, Linskens HF (1965) Protein diffusion from germinating pollen. Physiol Plant 18: 47-53

Stanley RG, Linskens HF (1974) Pollen - Biology Biochemistry Management. Springer Berlin Heidelberg New york

Stanley RG, Linskens HF (1985) Pollen - Biologie Biochemie Gewinnung und Verwendung. Urs Freund Greifenberg/Ammersee

Straka H (1975) Pollen - und Sporenkunde. Eine Einführung in die Palynologie. Grundbegriffe der modernen Biologie Band 13. G Fischer Stuttgart

Westerkamp C (1991) Honeybee are poor pollinators - why? Plant Syst Evol 177: 71-75

Wittig HJ, Welton WA, Burrell R (1970) A primer in immunological disorder. Thomas, Springfield

Wolff-Eisner A (1906) Das Heufieber, sein Wesen und seine Bedeutung. Lehmann, München

POLLEN FOR MONITORING ENVIRONMENTAL POLLUTION

M.J.M. Martens
Department of Biology
Faculty of Sciences
Catholic University
Toernooiveld 1
6525 ED Nijmegen
The Netherlands

Introduction

Pollen germination and pollen tube growth are affected by air pollutants *in vivo* and *in vitro*. A review of the effects on the different pollen parameters has been published earlier (Wolters and Martens, 1987). Since the reactions of plants to stress of any kind resemble each other, it is often difficult to distinguish the effects of air pollutants from the effects of other environmental factors. Moreover the response of plants to toxic compounds is also influenced by external and internal factors, such as temperature, relative humidity, stage of development of the plant and genetic resistance. The sexual organs can be affected directly by air pollutants. However, pollution-induced decrements of plant growth or foliar injury have an indirect effect on the generative parts of plants as well.

An other complication is that the degree of protection of pollen against contaminants alters during the different developmental stages during the sexual reproduction process. The pollen matures in the anthers, will be transferred to the air during distribution by wind or animals, will germinate on the stigma, and the pollen tube will grow in the style. The pollen susceptibility for air pollutants during these developmental stages can be studied *in vivo* by exposing pollen (1) in anthers, (2) during transfer, (3) on the stigma, or (4) during anthesis. Furtheron, the effect of pollution on the viability of pollen can be studied in long-term experiments, in which the plants or pollen are exposed to pollutants for a long time period. In that case also indirect influences of the sexual organs must be taken into account.

An alternative way is fumigation of pollen with toxic compounds in *in vitro* experiments. In literature three classes of exposure can be discriminated: (1) fumigation of dry pollen with dry gas at low relative humidity, (2) fumigation of dry or wet pollen with gas at high relative humidity, and (3) exposure of pollen in/on a germinating medium to gas or dissolved compounds.

The viability of pollen after or during exposure to air pollutants is studied both *in vivo* and *in vitro* concerning germination and tube elongation. The research up till 1985 concerning the response of pollen to (air) pollution is reviewed elsewere (Wolters and Martens, 1987, table I).

The effects of air pollutants on pollen

In the literature effects of air pollutants on pollen are described concerning gas exchange, chemistry, mitosis and DNA synthesis, morphology and weight, number of pollen grains per stigma, pollen-tube growth and germination. From the mentioned review (Wolters and Martens, 1987) the treatment of pollen in relation to the exposure of gases and heavy metals has been analysed. Table 1 offers a total result of this analysis. From this table it occurs that germination and pollen-tube growth are the two effects that are studied most in all reviewed literature during experimental exposure of pollen in *in vitro* experiments in germinating medium.

Table 1

Analysis of the treatment of pollen in relation to the exposure and the effect of the exposing treatment on pollen based on 65 references and 215 treatments described in these 65 references (from Wolters and Martens, 1987).

Exposure of pollen	Number of treatment
in vitro	
germinating medium	155
dry/wet pollen at high Rel. Hum.	14
dry pollen at low Rel. Hum.	32
on (fumigated) stigma	18
in vivo	
short-term during anthesis	
on (fumigated) stigma	7
in anthers	5
long-term, during development	30
Effects on pollen	
germination	141
pollen tube growth	135
number per stigma	5
morphology, weight	18
mitosis, DNA-synthesis	13
chemistry	26
gas exchange	1

Polluting substances studied are Sulphur dioxide, Nitrogen dioxide, Carbon dioxide, Carbon monoxide, Ozone, Ethylene, Formaldehyde, Acrolein, Aceton, Fluoride, Chloride, and Heavy metals as Cadmium, Chromium, Cobalt, Copper, Mercury, Lead, Nickel, Vanadium and Zinc. Often also the influence of mixtures of compounds have been investigated: Automobile exhaust gas, simulated Acid rain, Sulphur dioxide and Nitrogen dioxide, Sulphur dioxide and Formaldehyde, Nitrogen dioxide and Formaldehyde, Ozone and Sulphur dioxide, Ozone and Nitrogen dioxide, Ozone and Formaldehyde, Sulphur dioxide and Fluorides (Wolters and Martens, 1987). Also studied are the effects of Copper (Searcy and Macnair, 1990), detergents as Dodecylbenzen sulfonic acid (Paoletti, to be published) and Triton X-100 (Rao and Kristen 1990), Fungicides (Watters and Sturgeon, 1990), UV-B and Ozone (Feder and Shrier, 1990), Acid mist (Van Ryn et al., 1988). Phytotoxic compounds (AAL-toxins) inhibited *in vitro* pollen development from several susceptible *Lycopersicon* genotypes (Bino et al., 1988), which offers possibilities in plant breeding programmes and also an assay to distinguish between susceptible and resistant plants.

Cox (1988) studied the effects of combined simulated acis rain and Copper on the in vivo development of pollen (after *in vivo* pollination) related to various fruit and seed parameters and concluded that the inhibition of pollen on the stigma is responsible for the lack of seed set and therefore fruit abortion. Also fumigation with Sulphur dioxide of dry *Picea omorika* pollen resulted after pollination in reduced seed production (Krug, 1990).

Most of the information available on the impact of pollen concerns the effects of Sulphur dioxide. This is because (1) Sulphur dioxide is known to be phytotoxic over a long period, (2) it has been and is emitted in enormous amounts all over the world, and (3) it contributes to the formation of Acid rain. Because of their economic and esthetic value, crops and trees have been studied mostly concerning the interaction between air pollution and pollen.

Pollen as a Biomonitor

Biomonitors in a broad sense are organisms or communities of which the life-processes and the functioning are so tight correlated to specific environmental factors, that they can be used as indicators for these environmental factors. From this wide definition a specific concept of a biomonitor can be derived. Then a biomonitor is a biological substance, system or process that reacts to (a) specific time-dependant by man caused change(s) in an environmental factor or environmental factors, compared to a well defined normal reaction in an unpolluted situation (Schubert, 1991).

If pollen is to be used as a biological monitor for air pollution it is necessary to dispose of general conclusions and univocal relations between (concentrations of) pollutants and the caused response of the effect. In this contribution we will restrict ourselves to pollen germination and pollen tube growth as easy to study effects. From the literature it is clear that *in vitro* germination of pollen and pollen tube growth are much more sensitive to air pollutants than *in vivo* (Wolters and Martens, 1987). Nevertheless when pollen will germinate and grow in a germinating medium this can be compared to the situation on the surface of the stigma at high relative humidity, since in both situations air pollutants will dissolve in the water or in the moisture at the stigma respectively.

As has been pointed out most investigaters prefer the analysis of germination and pollen-tube growth in *in vitro* germination in a germinating medium. This way of handling seems therefore the best for using pollen as biomonitors.

Pollen can be used for monitoring if the pollen will be easily collectible and in large quantities. There should also be a continuous supply and the viability must be retained during storage. Pollen viability seems not to be affected by air contaminants during pollen transfer when dry, although we reported some effects in later developmental stages (Cox, 1988, Krug, 1990). Furtheron the germination percentage should be high and tube growth should be easily measurable. Because of these preconditions not all pollen will easily be used for monitoring or bio-assays. Stanley and Linskens (1985) have reviewed all factors, problems and possibilities in collecting, storing and using pollen from a lot of plant species for several purposes.

An optimal germinating medium should contain a carbohydrate (saccharose or glucose), Borium and Calcium, and further in several cases Magnesium and Kalium. The function of the carbohy- drate is twofold: it is a substrate for the germinating pollen and it functions as an osmoticum; the last function can be taken over by other osmotic substances that do not act as a substrate, for example polyetheleneglycol (Stanley and Linskens, 1985, page 76).

Pfahler (in press) summarizes the methods for growing Zea mays inflorescens (tassels) in order to be able to collect in an easy way maize-pollen that can be used for monitoring substances in the environment during *in vitro* germination on a semi-solid medium with bacto-agar or gelatin used as a solidifying medium, which is a precondition for oxygen exchange requirements. The surface should be kept in highly humid conditions te prevent desiccation. He also describes the use of *Nicotinana sylvestris* pollen as biomonitor in liquid germinating medium.

To prevent the time consuming microscopic scoring of the germina- ting percentage and measuring tube length a rapid and accurate photometric method has been developed by Kappler and Kristen (1987). After germinating the tobacco pollen in the culture medium containing the substances to be tested, the germinated pollen tubes were broken by sonication, purified and suspended in water. The optical density of this wall suspension measured at 500 nm was related to the amount of tube wall material which was produced within a defined period. By plotting the optical density values against the concentration of the polluting substances, signoid dose-response curves are to be obtained from which ED-50 values could be calculated. The optical density values versus time were directly proportional as well. Of course a zero-time control is necessary and was obtained from cultures killed with formaldehyde.

Concluding remarks

In vitro germination of pollen and pollen-tube growth in a liquified or a solidified germinating medium has occurred to be an easy to handle bio-assay for monitoring the effects of several substances and pollutants that can be present in the liquified or gaseous environment of the pollen. Much data available in the literature prove this conclusion. Therefore the pollen growth bio-assay is a cheap and well established test for investigating the effects of environmental pollutants and other agents on such a living system or biomonitor.

A problem that still exists at this moment is that as far as is known pollen of a specific species seems te be sensible to more than one agent. If this easy to handle bio-assay should be used to monitor unknown substances in the environment, research must be directed to finding specific and univocal relations between the effect of one substance and the reaction of (sensitive or non-sensitive) pollen of a specific plant species.

References

Bino RJ, Franken J, Witsenboer HMA, Hille J and Dons JJM (1988) Effects of *Alternaria alternaria* f.sp. *lycopersici* toxins on pollen. Theor Appl Genet 76: 204-208

Cox RM (1988) Sensitivity of forest plant reproduction to longe-range transported air pollutants: the effects of wet deposited acidity and copper on reproduction of *Populus tremuloides*. New Phytol 110: 33-38

Feder WA and Shrier R (1990) Combination of U.V.-B and ozone reduces pollen tube growth more than either stress alone. Env Exp Bot 30: 451-454

Kappler R and Kristen U (1987) Photometric quantification of *in vitro* pollen tube growth: a new method suited to determine the cytotoxicity of various environmental substances. Env and Exp Bot 27: 305-309

Krug E (1990) Reduced fertilization capacity of SO2-fumigated *Picea omorika*-pollen. Eur J For Path 20: 122-126

Paoletti E (to be published) Effects of acidity and detergent on *in vitro* pollen germination and tube growth in forest species.

Pfahler PL (in press) Analysis of Ecotoxic agents using Pollen tests. In: Linskens HF and Jackson JF (eds) Modern methods of plant analysis - New Series Vol 13: Plant toxin analysis. Springer Berlin Heidelberg New York London Paris Tokyo Hong Kong

Rao KS and Kristen U (1990) The influence of the detergent Triton X-100 on the growth and ultrastructure of tobacco pollen tubes. Can J Bot 68: 1131-1138

Schubert R (1991) Bioindikation in terrestrischen Oekosystemen. Gustav Fischer Jena, 2e ed.

Searcy KB and Macnair MR (1990) Differential seed production in *Mimulus guttatus* in response to increasing concentrations of copper in the pistil by pollen from copper tolerant and sensitive sources. Evolution 44: 1424-1435

Stanley RG and Linskens HF (1985) Pollen: Biologie, Biochemie, Gewinnung und Verwendung. Urs Freund Verlag, Greifenberg/ Ammersee

Van Ryn DM, Lassoie JP and Jacobson JS (1988) Effects of acid mist on *in vivo* pollen tube growth in red maple. Can J For Res 18: 1049-1052

Watters BS and Sturgeon SR (1990) The toxicity of some foliar nutrients and fungicides to apple pollen cv. Golden Delicious. Ann appl Biol 116(suppl): 70-71

Wolters JHB and Martens MJM (1987) Effects of air pollutants on pollen. Bot Rev 53: 372-414

SOUND PROPAGATION IN THE NATURAL ENVIRONMENT, ANIMAL ACOUSTIC COMMUNICATION AND POSSIBLE IMPACT FOR POLLINATION

M.J.M. Martens
Department of Biology
Faculty of Sciences
Catholic University
Toernooiveld 1
6525 ED Nijmegen
The Netherlands

Introduction

Acoustic communication by and between animals is dependent on the acoustic climate of the environment. Communication is ment to offer a method for sending a message from one animal to another, whether it is for occupation of a territory, attracting females for sexual reproduction as for example in the frog chorus, give warnings to other individuals of the same species in the case of detecting a predator, for detecting food or prey as for example the sonar system used by bats.

Sound propagation in the environment is a complex phenomenon influenced by the geometry of the sound source and the receiver, and the transmission path between both. This path is complicated by reflection and absorption of sound energy by the soil surface, reflection, absorption and scattering by branches, needles and leaves of the plants in the vegetation, and last but not least by the meteorological situation of the air volume between sound source and receiver. Moreover, it should be kept in mind that these factors influence each other and vice versa (figure 1).

Therefore an acoustic signal produced by an animal and passing the habitat will be degraded more or less by the acoustic climate, defined as the 'ambient sound' of that particular moment existing in the location being discussed (figure 2) and it depends on the actual environmental situation whether a signal will be received and the message understood by the receiving animal (Martens and Huisman, 1987).

The effects of environmental factors on sound propagation

An integral discussion concerning the effects of the soil surfaces and the micro- and macroclimate in the environment is offered elsewere (van der Heijden, 1984; Huisman, 1990). In figure 3 is shown what temperature gradients will be formed above a pasture, and in and above a forest during a sunny day. A real situation, measured above a hayfield, is shown in figure 4 together with the sound-ray-paths that will be produced at the temperature gradients

226

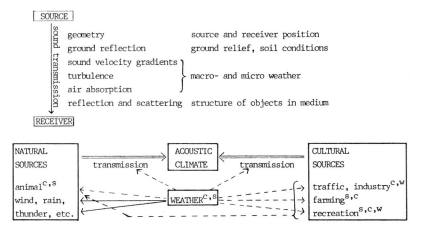

Fig.1 upper part. Scheme of factors influencing sound transmission. lower part. Scheme of factors influencing the Acoustic Climate; c= circadian, w=weakly, and s=seasonal rhythm.

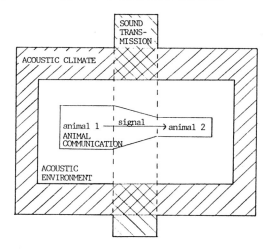

Fig.2 Relation between Acoustic Climate, Sound Transmission, Acoustic Environment and Animal Communication of a species.

and the sound levels that can be expected. A comparison of these model predictions with the actual measured sound levels revealed a close agreement (Huisman et al., 1987).

In a forested area measured temperature gradients are much more complicated as shown in figure 5 (Huisman, 1990). Here it is clear that vegetation has an important impact on the micro-climate and therefore also on the sound transmission path. Next to this effect the biomass in the vegetation causes absorption, reflection and scattering of sound as is shown in figure 6 (Martens, 1980) and in figure 7 (Martens, 1990).

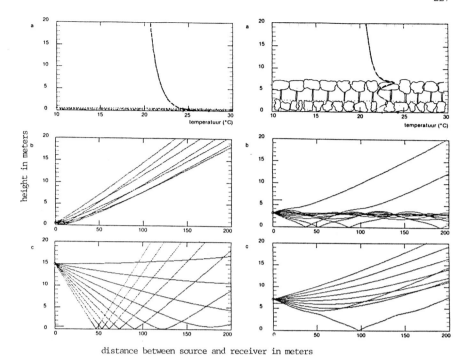

height in meters

distance between source and receiver in meters

Fig. 3 Temperature gradients during a sunny daytime and modelled sound transmission paths. Left hand side above a pasture: a. temperature profile, b. sound source close to the ground, note the sound shadow region, c. sound source located in the height. Right hand side above and in a closed forest: a. temperature profile, b. sound source in the canopy, note the tunneling effect just below the canopy, c. sound source above the canopy.

Acoustic communication by animals influenced by environmental sound propagation

In evolution animals have developed acoustic communication in such a way that interaction with the acoustic signals of other species can be avoided by temporal separation. Furthermore the size of the animal influences the height of the pitch and the sensitivity of the ear (Martens and Huisman, 1987), and the habitat structure has influenced the use of the acoustic vocabulary, which is clearly demonstrated to the differences in frequency and structure of acoustic signalling between forest birds and open field birds (Foppen and Martens, 1986). The moment of acoustic communication between individuals of the same species is often influenced by the 'acoustic climate' which is a result of the habitat structure and the by that habitat-structure caused micro-meteorological conditions. This is clearly shown by the dawn-chorus of long distance territory communication by birds and the night-chorus of frogs.

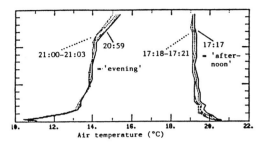

Fig. 4 Measured temperature profiles above a hayfield during afternoon and evening and modelled sound ray pattern over 40 m. $\triangle$ SIL is the measured in- or decrease compared to the median sound level in case of a neutral micro-meteorological atmosphere. Note the good agreement between modelled sound ray pattern and the measured sound levels at the different heights above the hayfield.

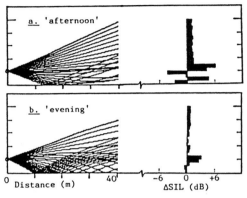

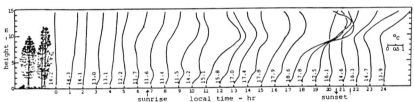

Fig. 5 Typical temperature profiles for a cloudy night, a clear but variable day and a very clear windless afternoon. Compiled from measurements of several days. Temperatures at height = 0.50 m are indicated.

Illustrative is the convergence and divergence in song features of a number of birds all over the world (Foppen and Martens 1986). In table 1 the frequency-characteristics of birds living in the tropical rain forest areas of Africa and Central America on the one hand and Europe on the other hand are compared. Differences are found between the tropical bird species living in the forests or in the open habitats, e.g. savannnas. Convergence is found within all forest-species versus open habitat-species. The same holds, but less clear, for European species living in different habitats. In table 2 the type of structure of songs is compared for the same groups. Birds communicating in forests mainly use pure tones, while birds in open habitats use thrills and sounds with a broad frequency range.

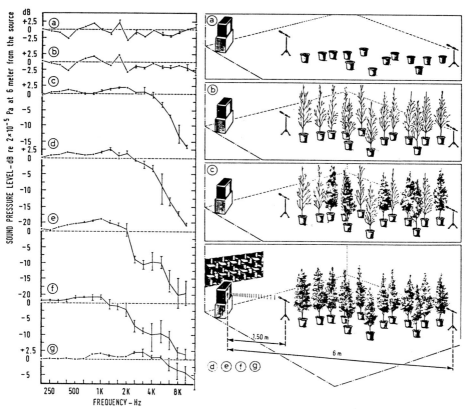

Fig. 6 Left hand side: Influence of 46 earthenware flowerpots on the sound field in the anechoic chamber; a. birch trees, all sawn down; b. all birch trees fully defoliated, stems, branches, and twigs present. Net influence of leaves, stems, branches, and twigs on the sound field in the anechoic chamber; c. 46 birch trees of which 23 are defoliated, the others foliated; d. 46 fully foliated birch trees; e. 25 fully foliated hazeltrees; f. 26 fully foliated tropical plants of different species; g. 12 fully foliated privets. The broken lines at 0 dB indicate: in a and b the sound field as it is generated by the sound source in the empty anechoic chamber at 6.0 m and put at 0 dB; in c to g the sound field generated in the anechoic chamber, when filled with the total number of empty flowerpots and with the trees sawn down, respectively. A dot in the curves indicates that the values of the sound pressure levels of three measurements were identical within 0.5 dB, and a vertical bar indicates the spreading of three measured values. Right hand side: Schematic view of the experimental arrangement in the anechoic chamber, corresponding to the curves at the left.

230

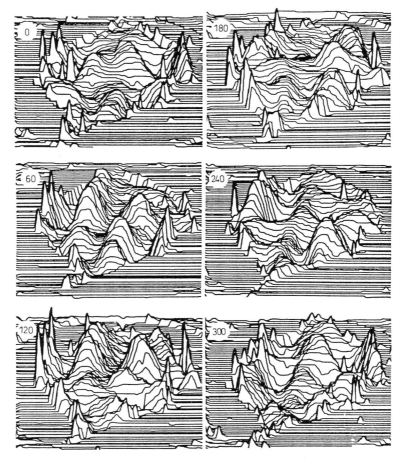

Fig. 7 Vibration velocity pattern of a free hanging *Euphorbia pulcherrima* leaf in a pure tone sound field of 400 Hz at 100 dB SPL measured by Laser-Doppler-Vibrometer-Scanning. From a full vibration cycle six regularly divided scans are pictured. The phase is indicated. The main nerve can be clearly recognized, and the surface and margins of the leaf show the largest velocity amplitudes.

The case of man is also illustrative for all other animals using acoustics for communication. Man has developed in forested areas. The acoustics of man is a clear result of the habitat situation. In that habitat a so cald 'sound window' can be found in de mid frequencies of sound. Therefore it can be understood that the frequencyspectrum produced by mans'voice fits in this sound window. Moreover the frequencyspectrum of our ears is most sensitive between 1 and 4 kHz, which is partly in that 'sound window' of forest acoustics.

table 1

habitat	Africa	Central America	Europe
forest	2.2 - 0.9 (25)		
low in forest		2.7 - 1.6 (19)	3.7 - 1.5 (10)
high in forest		3.6 - 1.8 (4)	4.0 - 1.1 (22)
half open	2.6 - 1.1 (11)	4.7 (34)	
open field	3.2 - 1.2 (24)	5.8 (6)	4.7 - 0.9 (16)

table 1.   Comparison of mean frequencies and standard deviations in the sound of song-birds living in different habitats in Africa (Chappuis 1971), Central America (Morton 1975) and Europe.   The number of species is given in between brackets.

table 2

habitat	type of structure of songs		
	pure tones	broad ranged	thrills
high in forest	74	26	21
low in forest	40	50	10
open field	18	81	56

table 2.   Structure of the song of Westeuropean song-birds living in different habitats.   Shown is the percentage of the total number of species investigated in a certain habitat having that peculiar structure.   Note that some bird-species can have different structures in their song.

Impacts for pollinating vectors

In the pollination syndrome vector-attraction by flowers is caused by offering the visiting animals food, whether pollen itself or nectar, warmth, as for example in *Trollius europeus* and many polar plant species, and pseudo-sex, as is found in many Orchid-species. Other proven attractants are the colour and odour of the flowers and the shape of the inflorescences.

From several insect eating bat-species it is known that they use ultra-sound (the sonar-system) to detect the prey.   Since it is known that plant organs can vibrate by the force of sound waves (Martens 1990) it could be interesting to speculate on the possibility that the sound waves produced by flying insects (e.g. the humming of bees and bumble-bees) might be reflected by the organs of flowers resulting in an acoustic gide to the flower and therefore an extra feature in the pollinating system.   No literature is available on this topic up to now, but it could be worthwhile to be investigated.

References

Chappuis C (1971) Un exemple de l'influence du milieu sur les emissions vocales des oiseaux: l'evolution des chants en forêt equatoriale. Terre et vie 25: 183-202

Foppen RPB and Martens MJM (1986) Investigating bird communication in relation to the acoustic environment. In: Martens MJM (ed) Proc Workshop on Sound propagation in forested areas and shelterbelts, Nijmegen Netherlands: 207-214

Heijden LAM van der (1984) The influence of vegetation on acoustic properties of soils. PhD Thesis, University of Nijmegen

Huisman WHT, Martens MJM and Asseldonk W van (1987) Measured and modelled temperature effects on outdoor sound transmission. Proc Inst Acoustics 9: 63-70

Huisman WHT (1990) Sound propagation over vegetation-covered ground. PhD Thesis, University of Nijmegen

Martens MJM (1980) Foliage as a low-pass filter: Experiments with model forests in an anechoic chamber. J Acoust Soc Amer 67: 66-72

Martens MJM (1990) Laser-Doppler vibrometer measurements of leaves. In: Linskens HF and Jackson JF (eds) Modern methods of plant analysis. New Series. Berlin Heidelberg New York London Paris Tokyo Hong Kong vol 11: 1-22

Martens MJM and Huisman WHT (1987) Sound propagation in the natural environment. In: Gelder JJ van, Strijbosch H and Bergers PJM (eds) Proc 4th Gen Meeting Soc Eur Herpetologica, Nijmegen Netherlands: 271-274

Morton ES (1975) Ecological sources of selection on avian sounds. Amer Natur 109: 17-34

CONFOCAL LASER SCANNING MICROSCOPY AND ITS APPLICATIONS IN RESEARCH ON SEXUAL PLANT REPRODUCTION

E.S. Pierson and A. Rennoch
Dipartimento di Biologia Ambientale
Università di Siena
Via P.A. Mattioli, 4
I-53100 Siena
Italy

Abbreviations: Charge-coupled device (CCD); Confocal Laser Scanning (CLS) Microscopy (M); Chlorotetracycline (CTC); Fluorescein-isothyocyanate (FITC); Photo Multiplier Tube (PMT); Tetramethyl rhodamine-isothiocyanate (TRITC).

Introduction

In the present chapter specific applications of CLSM in the research on sexual plant reproduction are briefly reviewed. The paper is addressed to students and university teachers of botany who are not familiar with CLSM, and includes introductory notes on the history and the principles of this technique (for authorized treatises on CLSM and fluorescence microscopy in biological sciences see: White et al., 1987; Pawley, 1990; Shotton, 1989; Wang and Taylor, 1989; Taylor and Wang, 1989).

History and principles of CLSM

In a conventional fluorescence microscope the full field in the specimen is illuminated (fig. 1a). By illuminating only a single specimen point at a time (fig. 1b), according to the "flying spot" illumination mode developed by Young and Roberts, the flare is reduced and the brightness of each specimen point can be quantitatively analyzed. Flare caused by unwanted light arising from different levels in the specimen can also be minimized by inserting a pinhole in the plane of focus, in front of a detector (fig. 2). One of the pioneers who considerably contributed to the application of these principles in microscopy techniques was Marvin Minsky, a postdoctoral fellow at Harvard University. In 1957, he applied for a patent for a microscope utilizing a stage-scanning optical system with one pinhole for limitation of the field of illumination, and

234

another, exit pinhole placed confocally to both the illuminated spot in the specimen and the first pinhole. Mojmir Petran and coworkers (Prague and New Haven) developed a "tandem" scanning confocal microscope, equipped with a scanner disc comprising multiple sets of holes, resembling a Nipkow disk. In 1972, Davidovits and Egger at Yale University obtained a patent for a laser illuminated confocal microscope (from Inoué, 1990).

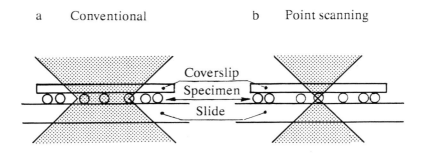

Fig. 1. Schematic representation of a) field illumination of the specimen as used in conventional microscopy, and b) point scanning illumination as currently applied in CLSM.

Fig. 2. Principles of confocal laser scanning fluorescence microscopy. A laser (La) beam is deviated to the microscope by a dichroic (Di) mirror. This beam (closed arrows) is further reflected by moving mirrors in a x-y scanning (Sc) pattern and focussed into the specimen (Sp) by an eye (E) lens and an objective (O) lens. The light beam induces excitation of fluorophores and autofluorescent particles within a certain depth of field in the specimen, and fluorescence light is emitted. The emission light beams (open arrows) initially follow the same way back as the excitation light beam (closed arrows). However, the emission light beams pass the dichroic mirror straight and are subsequently focussed by an additional lens (Le). Only the light emerging from the plane of focus in the specimen (—) passes the aperture (A) freely and reaches the detector (De), contributing to the imaging. Light which does not exactly stem from the focal plane (— — and ---) is suppressed at the level of A.

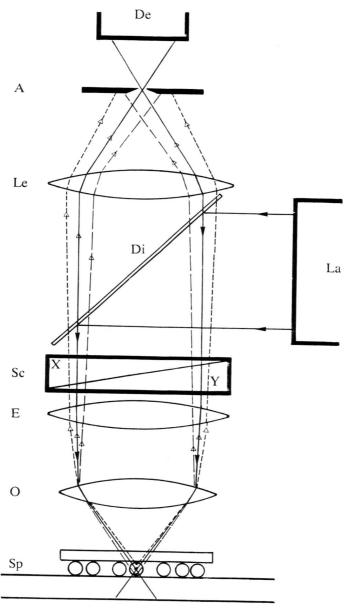

Fig.2

In the last five years, CLS microscopes have become current instruments in laboratories for the research of materials and biological specimens (White et al., 1987; Shotton, 1989). The confocal principle in epifluorescence scanning optical microscopy is illustrated in fig. 2 and briefly explained in the next paragraph.

While it is possible to apply a conventional xenon or a mercury lamp as a source for illumination in a confocal microscope, the laser is a preferable light source, due to its high degree of monochromaticity and high intensity of light. Commonly used in fluorescence microscopy are pure or mixed gas lasers, e.g. Argon, Krypton-Argon or Helium-Neon (0.3 mW) lasers. These lasers have single or multiline emission in the visible light range from 458 to 647 nm. The laser light beam is deviated to the microscope by a beam splitter or a dichroic mirror, the transmission characteristics of which depend on the wavelength of the light. The light beam is further reflected by moving mirrors and focussed by an objective lense into a very small spot in the specimen in a x-y scanning pattern. The plane of focus in z direction can be selected by operating a motor fixed on the micrometer screw, thus enabling "optical sectioning". Optical sectioning can be applied to a depth largely exceeding 100 micrometer in transparent tissues. Illumination of fluorophores or light scattering particles in the specimen results in the emission or reflection of a light beam which initially goes the same way back through the high numerical aperture objective as the exciting laser light beam. The emitted / reflected light beam passes the dichroic mirror / beam splitter and is focussed onto a pinhole (aperture). The light which does not stem exactly from the focal plane is effectively suppressed. Light from the focal plane in the specimen freely passes the pinhole and is detected by a PMT or in some apparatus by a CCD video camera. The size of the opening of the pinhole may be regulated. Maximal confocality is achieved with a nearly closed pinhole. The PMT or CCD converts the light signal into an electric signal, which is subsequently converted into a digital signal. The latter is displayed on a monitor in synchronism with the scanning laser beam in such a way that a two dimensional image of the scanned specimen is produced. Modern CLS microscopes are interfaced to a microcomputer with scan control, menu driven and image processing software, which usually consists of programs for image enhancements, arithmetic image operations, quantitative measurements, stereo and -moving- three dimensional reconstruction from optical sections and memory units for the storage of images (see for further details in Pawley, 1990).

In addition, in many CLS microscopes it is possible to conduct the laser light beam that has passed the specimen to a light detector and through glass fibers to a second PMT. The resulting image is a non-confocal transmission light image of the specimen. Double labeling imaging can be achieved by inserting appropriate filters in the way of the light beams so that fluorescence light of different wavelengths can be detected separately.

General evaluation of the CLSM technique

CLSM has a number of important advantages over conventional optical microscopy combined with direct photography using high speed films. The most important feature of confocal imaging is that out-of-focus flares, which often mask the relevant information, can be significantly reduced, and that the sensitivity is significantly higher. The horizontal resolution is better (under optimal conditions, lateral resolution about 200 nm and axial about 600 nm; Carlsson and Åslund, 1987). "Optical sectioning" is more accurate, and less laborious and disruptive compared to serial mechanical sectioning. The use of image enhancement functions is very convenient for improving the visualization of structures. However, elaboration of images involves the risk that relevant structures may be masked whereas others may be overestimated. Similarly, functions for the measurement of the size of structures and the pixel intensity ought to be used with care in studies aiming at semi-quantitative analysis, since the final image on the monitor depends on many factors, e.g. the laser type and laser power, the pinhole setting, the number of scannings and the quality of the optics. Moreover, some factors, such as the bleaching of the fluorochrome, the depth attenuation, the efficiency and stability of PMT and signal amplifiers are very difficult, if not impossible, to control. CLSM is not always the most appropriate approach to fluorescent imaging. Not all commercially available CLS microscopes are adapted for every fluorophore, especially not for those requiring ultraviolet excitation (Tsien and Waggoner, 1990). In some systems, the TRITC emission signal cannot be completely distinguished from the FITC signal. Rapidly moving particles in living cells, e.g. 3',3-dimethyloxacarbocyanine (DiOC) stained mitochondria, may be impossible to show with clarity by the normal relatively slowly scanning CLS microscopes. In these cases low light video imaging followed by image deconvolution (Agard et al., 1989) or tandem scanning confocal microscopy may give superior results. But in essence, the introduction of CLSM to fundamental and applied research and diagnosis has signified a real progress (review: Shotton, 1989).

Application of CLSM in research on sexual reproduction in Angiosperms

Only a few publications report the application of CLSM in the observation of reproductive organs of Angiosperms. Although these few papers do not cover the various aspects of sexual plant reproduction, they illustrate well the potency of the technique. The most relevant aspects of these works are resumed here below, ordered according to the type of imaging.

Reflection light and autofluorescence

To our knowledge, Fredrikson and co-workers (1988) were the first to use a CLSM for the study of the morphology of embryo sacs. They treated whole ovules of the orchid *Dactylorhiza maculata* with a clearing fluid (Herr, 1971) and observed in details the cell walls and the nuclei of young immature and fertilized embryo sacs by making longitudinal undistorted optical sections. They found that the pollen tubes start to grow before the female reproductive apparatus is mature. The pollen tubes seem to halt in the micropyle, waiting for the embryo sac's maturity. The pollen tube does not penetrate any of the synergids when entering the embryo sac, but becomes bladder-like and bursts, releasing the male cells. The megasporogenesis and fertilization process were also described in another species, *Herminium monorchis*, by following the same method (Fredrikson, 1990). In contrast to earlier reports using conventional approaches, the endosperm nucleus did not appear to be degenerated at the 3-celled stage.

In *Spinacia oleracea*, the autofluorescence displayed by the exine layer of the pollen grain has been used to demonstrate the hexagonal patterns formed by the germ pores by means of CLSM (Theunis et al, 1991). The autofluorescence property of the cell wall can be used to determine the approximate distance between a structure within the pollen grain and the pollen grain wall.

Recently, Reijnen et al. (1991) used CLSM for in situ hybridization in pollen of *Nicotiana tabacum* L. 'Petit Havanna'. Thick sections of pollen grains were incubated with the $[^{35}\text{-S}]$ and $[^3\text{-H}]$ labeled probes, synthesized from a pollen specific cDNA clone (pNTP303). Subsequently, the pollen samples were immersed in light sensitive emulsion and exposed in the dark for 10 or 20 days at 4° C, giving rise to the development of silver grains around the radioactive labeling. The reflectance of the silver grains, an indication of the hybridization signal, was analyzed by CLSM in optical sections at various levels of the pollen grain. Reijnen et al. (1991) concluded from their findings that the pollen specific RNA NTP303 was distributed over the whole vegetative cytoplasm of the pollen grain.

Fluorescent DNA staining

CLSM visualization of the detailed three-dimensional DNA structure of the pollen grain and pollen tube nuclei stained with 4'6-diamidino-2-phenyl-indole (DAPI) in *Galanthus nivalis* (Wilms et al., 1988) revealed that the generative cell and the vegetative nucleus, which are not connected in the mature pollen grain, become closely associated in the pollen tube grown in vitro. The association is still evident after 24 hours of growth. In *Spinacia oleracea*, the three-dimensional configuration of the trinucleate pollen grain was established by CLSM of the nuclei stained with ethydium bromide and propidium iodide (Theunis et al., 1991). The three nuclei are very close together, as in a male germ unit. They are positioned in the periphery of the pollen grain. In *Ornithogalum virens* (2 x 3x) propidium iodide staining has enabled the visualization of single chromosomes in dividing generative cells (fig. 3a).

Wagner et al. (1990) evaluated the surface area and volume of the vegetative and generative nucleus by measuring their periphery in consecutive optical sections. The authors found that both nuclei tended to enlarge during hydration.

Fluorescence labeling of the cytoskeleton

The cytoskeleton of male meiocytes, microspores, mature pollen or isolated embryo sacs (Huang et al., in progress) has been investigated by CLSM using rhodamine-phalloidin as a marker for F-actin, and monoclonal antibodies to reveal microtubules and a kinesin-like protein.

Very small amounts of rhodamine-phalloidin microinjected in growing pollen tubes of *Lilium longiflorum* were traced in the living vegetative cytoplasm by low light CLSM, revealing longitudinal filaments of actin (Zhang and Hepler, personal communication). In *Nicotiana tabacum*, serial optical sections show that numerous bundles of actin filaments envelop the generative cell in the pollen grain and that these bundles extend into the pollen tube (Derksen et al., personal communication). This finding may indicate that actin filaments are involved in the transport of the generative cell.

In metaphase I in male meiocytes of *Nicotiana tabacum*, opposite groups of microtubules, stained by an anti alfa tubulin, were observed to radiate from the poles and to form a distinct spindle (Tiezzi et al., 1991). The division of the generative cell in semi-vivo in *Nicotiana tabacum* (Bartalesi et al., 1991; fig. 3a) and in vitro in *Hyacinthus orientalis* (Tiezzi et al., 1989, Del Casino et al., accepted) is accompanied by spatial changes in the configuration of the microtubules, which are related to the arrangement of the chromosomes.

Theunis et al. (1992) found that the microtubular cytoskeleton of the generative cell, which initially has a basket-like configuration, desintegrates soon after isolation of the cell from the pollen grain.

Labeling with anti-kinesin reveals a granular pattern in the cortical region of the pollen tube tip in *Nicotiana tabacum* (conventional microscopy: Moscatelli et al., 1988; CLSM: Cai et al., 1992, Tiezzi et al., 1992b). The staining might correspond to Golgi-vesicles known to accumulate in this zone (for discussion on the possible meaning of this result see Cresti and Tiezzi, and Pierson and Li, this volume).

Fig.3a. Optical section through the micropylar part of a mature non-fertilized ovule of *Beta vulgaris* L. In this section, the egg cell (EC), a degenerated synergid (S), the micropyle (M) and the outer (OI) and inner integuments (II) are clearly visible. The ovule was fixed in 2 % paraformaldehyde and 2.5 % glutaraldehyde in 0.05 M phosphate buffer (pH 7.0), washed in buffer, dehydrated in 70 % ethanol and cleared with Herr's fluid (1971). The autofluorescence was detected by a PHOIBOS CLS microscope according to the procedure described by Fredrikson et al. (1988). (Magnification: 300x; courtesy of Dr. L. Bruun).
Fig. 3b. Organization of the actin skeleton in an isolated embryo sac of *Plumbago zeylanica*. Central cell nucleus (CCN), egg cell (E). Embryo sacs were isolated after enzymatic digestion of the ovules, essentially according to the method described in Huang et al. (1990; Amer J Bot 77: 1401-1410), fixed in paraformaldehyde, rinsed in buffer and stained with rhodamine-phalloidin. 25 direct images were taken at the same optical level, with a nearly open pinhole. These images were digitally "averaged", resulting in a final image with reduced background noise. (Magnification: 425x; Huang and Pierson, unpublished).
Fig. 3c. Bundles of microtubules in the generative cell of *Nicotiana tabacum*. Pollen tubes were grown in vitro for 3 hours, fixed in 3 % paraformaldehyde, washed in PIPES buffer, incubated with anti alfa tubulin (Amersham) and finally with a second antibody marked with FITC. Images were made with a Biorad MRC-500 CLSM equipped with a Nikon Optiphot microscope. The image is the result of a maximum projection of 9 subsequent optical sections taken with a step size of 1 micrometer. (Magnification: 3200x; courtesy of Dr. Li Yi-qin)
Fig. 3d. Single optical section showing the chromosomes stained with propidium iodide in a dividing generative cell of *Ornithogalum virens*, a 2 x 3x species.(Magnification: 2700x; Pierson, unpublished).

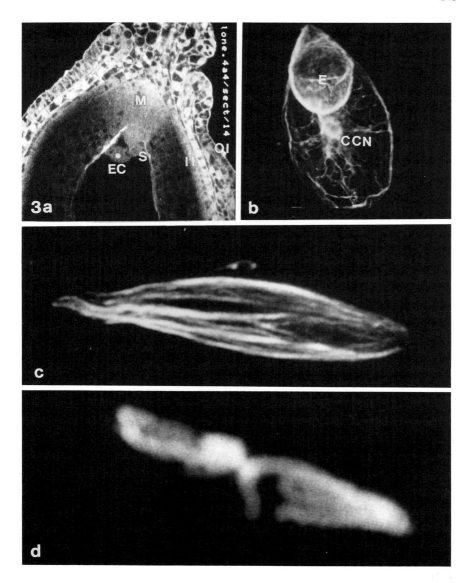

Fluorescent calcium probes

Many of the probes that are used to study free calcium or membrane associated calcium (Tirlapur, this volume) require to be excited by ultraviolet light, or light of very specific wavelengths, in the case of concentration measurements based on dual wavelength ratio imaging (critical review: Bolsover and Silver, 1991). The majority of the commercially produced CLS microscopes do not fulfil these conditions. By using another type of probe, Fluo-3-acetoxymethyl ester, Knebel and Reiss (1990) using a CLS microscope obtained indications on the distribution of cytosolic calcium in germinating pollen grains and pollen tubes of *Lilium longiflorum*. The image processing functions of a CLS microscope (Tirlapur and Cresti, communicated) were used to measure CTC fluorescence, an indicator for membrane associated calcium, on images recorded by conventional low light video microscopy in pollen tubes of Nicotiana tabacum. In growing pollen tubes a distinct tip to subapex gradient was observed, whereas no gradient was shown in the non growing tubes.

Prospects for the future

Nowadays, prototypes already exist of confocal scanning microscopes equipped with three or four different lasers to enable excitation in various wavelengths (e.g. Fricker and White, 1991). Multiple fluorescence confocal laser scanning imaging is thus a practical reality. The technique will certainly be extended to many fields of research, including sexual plant reproduction. It is expected that further application of CLSM for spatial localization of epitopic sites, mRNAs and portions of the DNA genome will further enable the construction of three-dimensional maps of pollen, embryo sacs and other tissues of flowering parts of Angiosperms. Probably CLSM detection of fluorescent in situ hybridization (e.g. Brakenhoff et al., 1990) will be introduced to the study of reproductive organs of plants very soon. A combination of CLSM with supravital staining (Tsien and Waggoner, 1990) and other fluorescent compounds that can be microinjected (like conjugated antibodies, cytochemical analogs, markers and drugs; Wang and Taylor, 1989) opens wide prospects for the investigation of the cellular biology of living cells. CLSM is thus making an impressive contribution to the present revival of optical microscopy.

Acknowledgements

The preparation of this article was supported by grants from Consorzio Siena Ricerche (to AR) and the European Community (to ESP, BAP-0597-I-CH and BIOT-0078-I-CH). Dr. Li Yi-qin is gratefully acknowledged for critically reading the manuscript and R. Pierson for making the drawings.

References

Agard DA, Hiraoka Y, Shaw P, Sedat JW (1989) Fluorescence microscopy in three dimensions. In: Fluorescence microscopy of living cells in culture. Part B. Quantitative fluorescence microscopy- Imaging and spectroscopy. Taylor DL, Wang YL (eds). Academic Press San Diego New York Boston London Sydney Tokyo Toronto pp. 353-377

Bartalesi A, Del Casino C, Moscatelli A, Cai G, Tiezzi A (1991) Confocal laser scanning microscopy of the microtubular system of dividing generative cell in *Nicotiana tabacum* pollen tube. Gior Bot Ital 125: 21-28

Bolsover S, Silver RA (1991) Artifacts in calcium measurement: recognition and remedies. Trends Cell Biol 1: 71-74

Brakenhoff GJ, Visscher K, Van der Voort HTM (1990) Size ans shape of the confocal spot: control and relation to 3D imaging and image processing. In: Handbook of biological confocal microscopy. Pawley JB (ed) Plenum Press New York London pp.87-91

Cai G, Bartalesi A, Moscatelli A, Del Casino C, Tiezzi A, Cresti M (1992) Microtubular motors in the pollen tube of *Nicotiana tabacum*. Proc Int Symp on Angiosperm Pollen and Ovules: Basic and Aspects. Villa Olmo, Como June 23-27 1991 (in press)

Carlsson K, Aslund N (1987) Confocal imaging for 3-D digital microscopy. Appl Opt 26: 3232-3238

Del Casino C, Tiezzi A, Wagner VT, Cresti M (accepted) The organization of the cytoskeleton in the generative cell and sperms of *Hyacinthus orientalis* L. Protoplasma

Fredrikson M (1990) Embryological study of *Herminium monorchis* (Orchidaceae) using confocal scanning laser microscopy. Amer J Bot 77: 123-127

Fredrikson M, Carlsson K, Franksson O (1988) Confocal scanning laser microscopy, a new technique used in an embryological study of *Dactylorhiza maculata* (Orchidaceae). Nord J Bot 8: 369-374

Fricker MD, White NS (1991) Multi-wavelength techniques in confocal scanning laser microscopy. Int Bot Microsc Meeting. Durham University. 8- 12 April 1991 Abstract

Herr JM (1971) A new clearing squash technique for the study of ovule development in angiosperms. Amer J Bot 58: 785-790

Inoué S (1990) Foundations of confocal scanned imaging in light microscopy. In: Pawley JB (ed). Handbook of biological confocal microscopy. Plenum Press New York London pp. 1-14

Knebel W, Reiss HD (1990) Visualization of cytosolic Ca^{2+}-distribution with the confocal laser scanning microscope (CLSM) of Leica Lasertechnik. Sci Techn Information Leica 9: 279-283

Pawley JB (1990) Handbook of biological confocal microscopy. Plenum Press New York London

Reijnen WH, van Herpen MMA, de Groot PFM, Olmedilla A, Schrauwen JAM, Weterings KAP, Wullems GJ (1991) Cellular localization of a pollen-specific mRNA by in situ hybridization and confocal laser scanning microscopy. Sex Plant Reprod 4: 254-257

Shotton DM (1989) Confocal scanning optical microscopy and its applications for biological specimens J Cell Sci 94: 175-206

Taylor DL, Wang YL (eds) (1989) Fluorescence microscopy of living cells in culture. Part B. Quantitative fluorescence microscopy- Imaging and spectroscopy. Methods in Cell Biology Vol 30 Academic Press San Diego New York Boston London Sydney Tokyo Toronto pp. 353-377

Theunis CH, Cresti M, Milanesi C (1991) Studies of the mature pollen of *Spinacia oleracea* after freeze substitution and observed with confocal laser scanning fluorescence microscopy. Bot Acta 104: 324-329

244

Theunis CH, Pierson ES, Cresti M (1992) The microtubule cytoskeleton and the rounding of isolated generative cells of *Nicotiana tabacum*. Sex Plant Reprod 4 (in press)

Tiezzi A, Moscatelli A, Murgia M, Russell SD, Del Casino C, Bartalesi A, Cresti M (1989) Immunofluorescence studies on microtubules in the male gamete of *Hyacinthus orientalis* and *Nicotiana tabacum* using confocal scanning laser microscopy. In: Proc Third Sperm Cell Meeting. Barnabas B, Liszt K (eds), Martonvasar pp. 17-21

Tiezzi A, Bednara J, Del Casino C, Bartalesi A, Cai G, Moscatelli A (1992a) The microtubular cytoskeleton during pollen development and pollen tube growth in *Nicotiana tabacum*. Proc. Int. Symp. on Angiosperm Pollen and Ovules: Basic and Applied Aspects. Como June 23-27, 1991 (in press)

Tiezzi A, Moscatelli A, Cai G, Bartalesi A, Cresti M (1992b) An immunoreactive homolog of mammalian kinesin in *Nicotiana tabacum* pollen tubes. Cell Motility Cytoskeleton (accepted)

Tsien RY, Waggoner A (1988) Fluorophores for confocal microscopy: photophysics and photochemistry. In: Pawley JB (ed). Handbook of biological confocal microscopy. Plenum Press New York London   .

Wagner VT, Cresti M, Salvatici P, Tiezzi A (1990) Changes in volume, surface area, and frequency of nuclear pores on the vegetative nucleus of tobacco pollen in fresh, hydrated and activated conditions. Planta 181: 304-309

Wang YL, Taylor DL (eds) (1989) Fluorescence microscopy of living cells in culture. Part A. Fluorescent analogs, labeling cells, and basic microscopy. Methods in Cell Biology Vol 29. Academic Press San Diego New York Boston London Sydney Tokyo Toronto

Wilms HJ, Murgia M, Van Spronsen (1988) Confocal scanning laser microscopy of *Galanthus* generative cells. In: Wilms HJ, Keijzer CJ (eds) Plant sperm cells as tools for biotechnology. Pudoc Wageningen

Author index

Subject Index

Printing: Mercedesdruck, Berlin
Binding: Buchbinderei Lüderitz & Bauer, Berlin

DATE DUE

AUG 2 5 1994	
AUG 1 0 1994	